The Great Comet Crash

The Great Comet Crash

The impact of Comet Shoemaker–Levy 9 on Jupiter

EDITED BY

John R. Spencer
Lowell Observatory

and Jacqueline Mitton

Published by the Press Syndicate of the University of Cambridge
The Pitt Building, Trumpington Street, Cambridge CB2 1RP
40 West 20th Street, New York, NY 10011-4211, USA
10 Stamford Road, Oakleigh, Melbourne 3166, Australia

First published 1995

Printed in Great Britain at the University Press, Cambridge

A catalogue record for this book is available from the British Library

Library of Congress cataloguing in publication data available

ISBN 0 521 48274 7 hardback

Contents

Editors' preface

This book relates the remarkable story of comet Shoemaker–Levy 9, from its discovery in March 1993 to its spectacular demise in Jupiter's atmosphere in July 1994. The tale is told in the words of some of the astronomers and planetary scientists who played key roles in the drama, and also by means of numerous images and other illustrations, for this is a visually dramatic story. We have kept the technical jargon to a minimum (and have added a glossary to explain those technical terms that we could not avoid), but have included enough scientific detail to explain not just what happened, but why it happened, to the extent that this was understood in the Spring of 1995. No doubt some of what we say will be superceded in the months and years to come, but we hope (and expect) that the overall picture that we paint will continue to be valid.

The astronomical community responded to the comet collision with an unprecedented show of global cooperation and generosity in the sharing of observations and ideas, and we were delighted to find that the spirit of generosity continued as we gathered material for this book. We extend our heartfelt thanks to everyone who contributed images and other data, and answered our many questions, and to all the other people who made this book possible by their remarkably successful, and frequently heroic, efforts to predict, record, and understand the spectacle. We also apologize to all those who sent us material that we could not include. Particular thanks are due to Tom Herbst, Heidi Hammel, Ann Sprague, Paul Chodas, John Trauger, Phil Nicholson, Tim Livengood, Charlie Avis and Imke de Pater. The collection of images put together on CD-ROM by Network Cybernetics Corporation also proved invaluable.

We would like to acknowledge the major role played by Simon Mitton of Cambridge University Press in initiating this project and supporting us with practical assistance at every stage of bringing the book together.

John Spencer thanks his wife Jane for her loving support, valued advice, and her understanding during all those weekends spent in front of the computer.

John R. Spencer, Flagstaff, Arizona, USA
Jacqueline Mitton, Cambridge, UK

MAY 1995

Foreword

Gene and Carolyn Shoemaker

The crash of some 20 nuclei of Periodic Comet Shoemaker–Levy 9 into Jupiter between 16 and 22 July 1994 provided astronomers and other scientists with a unique chance to examine directly one of the most fundamental processes in the evolution of our solar system.

It is now widely accepted that the terrestrial planets, Mercury, Venus, Earth, and Mars, as well as the two outer giant planets, Uranus and Neptune, were formed by the accumulation of small bodies through collision. Moreover, the cores of Jupiter and Saturn probably were formed this way, and the accretion of cometary matter over the course of solar system history probably has contributed significantly to the atmospheres of the planets. It is very likely that most of the water, carbon, and nitrogen on Earth was brought by the impact of comets. Besides providing the environment necessary for life, these materials are basic constituents of living organisms.

The chemical composition of typical comets, according to our present concept, is rather similar to the composition of the human body. The principal difference is that most comets probably contain iron and magnesium silicates. These 'rocky' components are not found in most living organisms. Comets also contain a variety of organic compounds in the form of solid grains (the so-called CHON particles detected in the coma of Comet Halley), which may include amino acids, the building blocks of protein and living tissue. Some scientists have speculated that organic compounds delivered to Earth in the form of cometary dust may have provided the starting material from which life arose. Alternatively, complex chemical reactions occurring in the fireballs produced by comet impacts may have been important sources of the chemicals necessary for biological evolution to start.

Over the past decade, many scientists from diverse disciplines have focused on the possibility that the history of life on Earth has been punctuated by mass extinctions of species, following global environmental catastrophes produced by the impacts of interplanetary bodies. There are now fairly strong reasons for thinking that the body that struck Earth 65 million years ago, at the time of the great mass extinction at the end of the Mesozoic Era, was a comet. It produced a giant crater, at least 170 km in diameter, that now lies buried beneath a kilometer of marine limestone on the northern tip of Yucatan in Mexico. Possibly this was one of the most energetic impact events on Earth over the past half billion years, but the geologic record of impact craters is so incomplete we cannot tell. Circumstantial evidence links at least four other mass extinctions to large impacts or to sequences of large impacts.

From the standpoint of the history of life, the most interesting consequence of the mass extinctions is not so much the loss of certain species but the blossoming of other species afterwards. It can plausibly be argued not only that we are the direct descendants of cometary material, but that the rise of our species is one ultimate consequence of the rapid increase in the variety of mammals inhabiting Earth following the mass extinction of 65 million years ago. The proposition, in other words, is that we are the progeny of comets and our own species has arisen because the course of evolution was fundamentally altered by a comet impact.

The recognition of impact catastrophes in the geologic past has also spurred examination of the hazard to humankind or to human civilization posed by large impacts in the future. There is no question that rare, large, impact events can produce acute global alterations of the environment that surpass other known natural hazards. The magnitude of the risk, however, depends very sensitively on the minimum size of impacting body capable of producing a substantial change in the global environment. Generally, the most serious short-term environmental effects are thought to arise when light-absorbing dust and aerosols become suspended in the atmosphere. The threshold diameter for a body that can cause such effects has been suspected to be about 2 km. The average time between impacts of bodies this large is about half a million years, an interval vastly longer than the history of civilization. But the threshold size is, in reality, unknown and may be very different according to whether the object is a comet or an asteroid. The observations of the dark clouds created by the impact of SL9 on Jupiter lead us to think that global catastrophes on Earth might be caused by bodies less than 2 km across and might be far more frequent than previously supposed.

For all of these reasons, and others, the 'crash of '94' on Jupiter seized the imagination of scientists and lay persons alike. For the first time, we had the chance to witness a sequence of high speed impacts that delivered to the atmosphere of Jupiter a total amount of energy perhaps one to two orders of magnitude less than the energy released in the giant impact on Yucatan. Impacts on this scale could previously be studied only by complex modeling with numerical codes on high-speed computers. Nature could now provide a reality check for all the assumptions and for the reliability of those codes.

But beyond the physics of impacts, there were major questions about the composition and physical structure of cometary nuclei, the chemistry of the hidden layers of the jovian atmosphere, and the chemical reactions that might take place as the fireballs cooled down from temperatures of tens of thousands of degrees. Beyond the more obvious questions was the prospect of discovering new or wholly unanticipated phenomena. What would happen to the jovian magnetosphere? Would we see waves propagated in the atmosphere? Would we learn something about the circulation pattern in the largely invisible stratosphere?

Circumstances seemed to conspire to provide the best opportunity in a century, perhaps in a millennium or more, to observe the impact of a comet on a planet. The comet had passed within a third of a jovian radius above Jupiter's cloud tops in 1992. It was torn asunder by tidal forces at that time. Swarms of fine debris were generated, and the surfaces of the larger fragments were newly exposed to sunlight. Very likely, more dust was released by rejuvenated cometary activity – the slow sublimation

of ice freshly exposed to the Sun. Thus the comet, which was previously too faint to be detected photographically with wide-field telescopes, suddenly became detectable because of the sunlight reflected from the dust.

The fortunate discovery of SL9 sixteen months before it would hit Jupiter, and the recognition two months later by Brian Marsden and Syuichi Nakano that the comet probably would strike Jupiter, gave astronomers ample opportunity to request substantial periods of observing time on many of the largest optical telescopes at major observatories around the world. It also gave funding agencies time to organize and provide financial support to investigators. There was time to consider observing plans carefully and to bring into play state-of-the-art equipment. Had the impacts occurred a decade earlier, astronomers would not have had at their command the infrared imaging systems that proved so valuable in recording the fireballs and their evolution into warm clouds. Fortuitously, the successful mission to repair the Hubble Space Telescope was completed 7 months before the crash. With its performance improved dramatically, the HST was available for detailed studies of the comet itself, individual impact events, and the later evolution of the impact plumes.

The impacts might have occurred when Jupiter was close to conjunction with the Sun and unobservable from Earth, but Jupiter was fairly well placed in the sky. They might have occurred near the middle of the far side of Jupiter, as seen from Earth. Instead, all the impacts took place slightly beyond the limb of Jupiter, as seen from Earth. The *Galileo* spacecraft, en route to Jupiter, was in a position to observe them directly. Furthermore, the break-up of the original comet produced multiple nuclei with a correspondingly large number of opportunities to observe individual impact events over the course of five and a half days.

Theorists bent to the task of calculating the possible effects of the impacts. Early on, it was realized that plumes of hot gas from the fireballs created would rise very rapidly and should be detectable from Earth. In fact, excellent observations were obtained of the vertical profiles of the rising plumes. The rotation of Jupiter soon brought them fully into view.

There were, of course, the doubters. Many thought the comet nuclei would disappear without a trace. The impacts would be a 'big fizzle'. Even the most optimistic participants, including ourselves, worried that the effects might be difficult to observe.

We were all wrong. The phenomena were spectacular, and most observations obtained exceeded expectations. The plumes radiated copious amounts of energy at infrared wavelengths. A wide variety of chemical compounds were detected spectroscopically in the plumes, some quite unexpected. Radio emission from the magnetosphere increased substantially during impact week. One wholly unanticipated phenomenon, observed 45 minutes after the impact of one of the brightest nuclei, was the release of charged particles from the magnetosphere, which produced new auroral spots in both the northern and southern hemispheres. Low velocity waves expanding outward from some of the crash sites were seen in HST images somewhat mysteriously as dark rings in the stratosphere.

To everyone's surprise, many impacts produced huge dark clouds that were very easily detected at visible as well as infrared wavelengths. Even amateurs with small telescopes could see and follow the evolution of the dark clouds. And highly skilled amateurs were ready, in force. CCD images obtained by amateurs with moderate-size telescopes may constitute the most complete record of the long-term evolution of the impact clouds. Determination of the optical depth of the clouds (their capacity to absorb sunlight) and observations of their duration may have the most direct relevance to our understanding of the effects of comet impacts on Earth. From its color, we can infer that the dark material is composed chiefly of organic compounds. Much of the carbon may have come from the comet. Even as we write, a band of dark clouds is still visible in Jupiter's southern hemisphere. How long will it last?

This book is the first, general, profusely illustrated overview of the comet crash to be published in English. We hope that it reaches a broad audience. In David Levy's words, 'Comet Shoemaker–Levy 9 was everybody's comet'. Its collision with Jupiter was an event that raised world-wide awareness of the dynamic solar system in which we reside.

December 1994

1

Introduction

Jacqueline Mitton

Comets are coming and going all the time. Occasionally one falls into Jupiter's gravity trap. A few comets break up into pieces. But the amazing story of the life and death of Comet Shoemaker–Levy 9 is unique in recorded astronomical history.

This book is about a drama, played out against the backdrop of the starry sky between two principal characters, a comet and a planet. An Earth-bound audience of millions watched in awe as a broken comet inexorably hurtled along a path to total destruction. When the end came, the death throes of the comet were a truly dramatic climax and the planet was scarred by wounds inflicted in the fearsome encounter. Astronomers had never seen anything remotely like it before.

Such violence unleashed in our own back yard (astronomically speaking), struck hard at human perceptions of Earth as a safe haven of life. No-one who understood what happened could be a dispassionate spectator. Expiring gloriously, Comet Shoemaker–Levy 9 (SL9) won a special place in the annals of science and left a permanent impression of our own fragility in the face of cosmic collisions.

Yet by some miracle we were here in 1994 to see the Great Comet Crash, able to predict what would happen to the comet 16 months after its discovery and to make a shrewd guess at what it had been doing for the previous few decades. Now theorists have inherited a unique record of a whole series of flashes, plumes, and resulting dark clouds to interpret. There are more clues as to the true characters of our actors, but can their innermost secrets be revealed?

What are comets?

A comet visible to naked-eye observers appears every few years on average. In almost every case, they arrive unpredicted. Several dozen different comets are seen by astronomers at some time or another over the course of a year, but many are too faint for telescopes used by amateurs. No two comets ever look identical, but they have basic features in common. There is a coma, which looks like a misty patch of light. One or more tails often stream from the coma, in the direction away from the Sun.

At the heart of a comet's coma lies a nucleus of solid material, typically no more than 10 km across. The visible coma is a huge cloud of gas and dust that has escaped from the nucleus, which it then surrounds like an extended atmosphere. The coma can extend as far as a million kilometers. Around the coma there is often an even larger invisible envelope of hydrogen gas.

The most graphic proof that the grand spectacle of a comet develops from a relatively small and inconspicuous chunk of ice and dust was the close-up image obtained in 1986 by the European *Giotto* probe of the nucleus of Comet Halley. It turned out to be a bit like a very dark asteroid, measuring 16 by 8 km. Ices have evaporated from its outer layers to leave a crust of nearly

In March 1986, the ESA spacecraft *Giotto* approached within 2000 km of the nucleus of Halley's comet. This is a composite of six images taken at distances between 14 400 and 2700 km. The black nucleus is 15 km long and 8 km wide. On the left, jets of gas and dust spurt from active holes in its surface and are illuminated by sunlight. (H. U. Keller/Max-Planck-Institut für Aeronomie)

black dust all over the surface. Bright jets of gas from evaporating ice burst out on the side facing the Sun, where the surface gets heated up, carrying dust with them. This is how the coma and the tails are created.

Comets grow tails only when they get warm enough for ice and dust to boil off. As a comet's orbit brings it closer to the Sun first the coma grows, then two distinct tails usually form. One, the less common kind, contains electrically charged (i.e. ionized) atoms of gas, which are blown off directly in the direction away from the Sun by the magnetic field of the solar wind. The other tail is made of neutral dust particles, which get gently pushed back by the pressure of sunlight itself. Unlike the ion tail, which is straight, the dust tail becomes curved as the particles follow their own orbits round the Sun. The gas in the ion tail can absorb solar energy and then emit it again, but the dust tail shines only by reflecting sunlight.

For all their size and brightness, comets contain very little material, perhaps around one billionth the mass of Earth. During a passage around the Sun, well under one percent of that material is typically lost to make the coma and tails.

At Jupiter's distance from the Sun, Comet Shoemaker–Levy 9 was too cold for much ice to evaporate and would have remained relatively inactive and undetectable had it not been shattered. When it was torn apart, quantities of dust were released to create comae and tails for the 20 or so individual fragments, which turned into a string of mini-comets. With plenty of dust to reflect sunlight, SL9 became bright enough to be discovered by astronomers on Earth.

Where do comets come from?

Comets must come from somewhere, because we keep discovering new ones that have arrived close enough to Earth and the Sun to be visible. One likely source is a vast reservoir of dusty ice balls surrounding the solar system at a distance of order one light year, a thousand times further away than the planet Pluto. The Dutch astronomer Jan Oort developed this theory in the 1950s. The presumed shell of a thousand billion potential comets is known as the Oort cloud. Oort cloud comets were probably formed much closer to the Sun but were ejected into their distant orbits by the gravity of the giant planets, early in the history of the solar system. A small proportion of its members enter the inner solar system when they receive gravitational kicks in our direction from passing stars.

Other potential comets may be stored rather closer, in the region of the solar system just beyond Neptune. Since 1992, when the first faint, small body was discovered, the count of known objects orbiting there has steadily been rising by several a year. There are almost certainly thousands awaiting discovery. This ring of debris left over from the formation of the solar system is known as the Kuiper belt (or disk) after Gerard P. Kuiper, whose theoretical work on the origin of the solar system predicted its existence.

Jupiter, Master of Comets

Jupiter, the most massive of the planets, exerts a highly significant gravitational influence in the solar system. A mere comet nucleus easily falls under Jupiter's power if it approaches too closely. The usual outcome of such an encounter is that the comet is ejected from the solar system, or is perturbed into a new elliptical orbit around the Sun. In this new orbit, its maximum distance from the Sun becomes about the same as Jupiter's, and its period is typically around six or eight years. There are about 50

A typical comet in orbit around the Sun grows its tail only when it is in the inner solar system. Gas streams directly away from the Sun in a straight line, while the dust develops into a broad, curved tail. (From *The Cambridge Atlas of Astronomy*, courtesy Encyclopaedia Universalis)

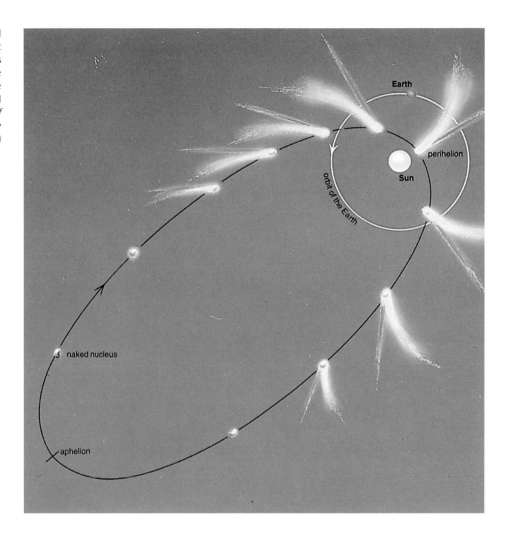

known comets trapped into orbits like this. Collectively, they are known as Jupiter-family comets.

In the case of SL9, the interaction was even more intimate. This comet was pulled into orbit around Jupiter itself. It was a loosely-bound orbit, but nevertheless Jupiter managed to hang on in the tug-of-war with the Sun for control. A handful of comets have fleetingly been temporary satellites of Jupiter before, but never has one survived in jovian orbit so long. At least there have not been any we could see. Quite possibly they are or have been there, but as long as they remain intact and inactive, we are most unlikely to detect them.

Long and short period comets

Astronomers draw a distinction between comets trapped in elliptical orbits within the planetary system, which are seen time and again, and those that shoot by the Sun's immediate vicinity with no evidence that they can return in under 200 years. The transients are called 'long period comets'. The regular returners are called 'short period' or sometimes just 'periodic' comets and their names, when written in full, are prefixed by 'P/'.

As a captured comet, though orbiting Jupiter rather than the Sun, Comet P/Shoemaker–Levy 9 was dignified with the periodic prefix in its official designation. The Shoemaker/Levy team had previously discovered more than nine comets, but this was the ninth *periodic* comet.

Disintegration of comets

The break-up of comets is not unusual. In the records, there are more than 20 that have split into two or more parts. The material a cometary nucleus is made of is not particularly strong and any kind of stress, or just the evaporation of the ice that holds it together, could have a

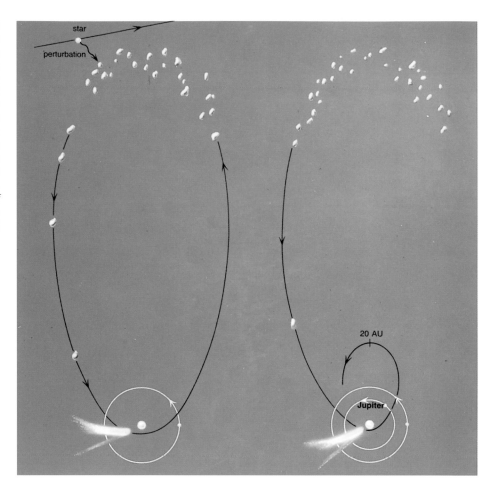

A reservoir of comet nuclei surrounding the solar system at about a light year from the Sun is a likely source of many of the comets that eventually arrive in the inner solar system. A gravitational disturbance from a passing star could be enough to catapult a potential comet in our direction (left). If the comet then passes close to Jupiter (right) its path can be changed again so that it becomes a short-period comet. (From *The Cambridge Atlas of Astronomy*, courtesy Encyclopaedia Universalis)

destructive effect. Falling apart may simply be part of the ageing process for normal comets.

The classic prototype is Comet Biela. When W. van Biela found it in 1826, he was able to calculate its orbit accurately enough to identify two previously recorded apparitions. At a return in 1846 it had split in two. In 1852, the two parts were observed following the same orbit, but more than two million kilometers apart. It was never seen again. A more recent example was Comet West of 1976. Its nucleus split into four parts during March of that year. But since it is not a short period comet, we shall never know their ultimate fate.

In the case of SL9, however, the cause of disintegration was patently clear. When the comet passed very close to Jupiter in 1992, the gravitational pull on its nucleus was slightly stronger on the side closest the giant planet than on the opposite side. This slight force imbalance, known

as a tidal force, was sufficient to tear the comet apart. A similar fate overtook Comet P/Brooks 2, discovered by William R. Brooks in 1889. Three years earlier it skimmed Jupiter and three pieces broke off. It survived, but clearly in a weaker state. It has faded and is at least a thousand times fainter than it was 1889. The 1886 encounter flung it into an entirely new orbit and changed its period to 7 years from 29 years.

Tidal forces close to the Sun have sometimes caused comets to break up. The Great Comet of 1882 ploughed through the Sun's outer layers, only one third of a solar radius from the level we see as the 'surface'. Afterwards, the nucleus was elongated and contained several bead-like condensations. In 1979, astronomers working on an orbiting solar observatory (called SOLWIND) discovered a comet already deep in the Sun's corona, which plunged into the Sun.

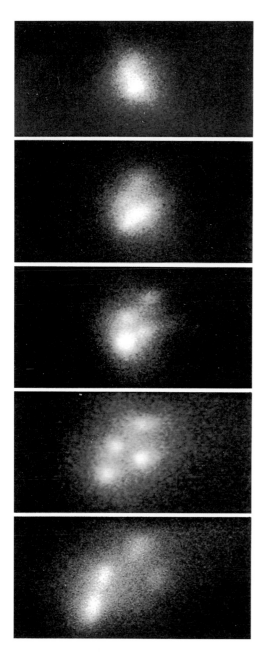

The nucleus of Comet West, which appeared in 1976, broke up spontaneously into four pieces of similar size. Each later developed a separate tail. These images were taken on 8, 12, 14, 18 and 24 March 1976. (C. F. Knuckles & A. S. Murrell, New Mexico State University)

Images across the spectrum

This book is no ordinary photograph album. Many of the astronomical images in it differ from conventional photographs in both the type of light they record and the technology used to create them.

Our familiar view of Jupiter is in visible light – the kind of radiation our eyes are tuned to detect. Visible light is composed of electromagnetic waves with a wavelength of about 500 nanometers. (A nanometer, abbreviated to nm, is one millionth of a millimeter.) Jupiter shines in the night sky with sunlight reflected from its clouds. The huge dark marks seen around many of the crash sites were material that doesn't reflect sunlight so well as the usual clouds.

The comet impacts, however, had effects that could be seen by viewing radiation over a huge range of wavelengths, both longer and shorter than visible light. Many of the professional observations were made with detectors sensitive to radiation bands invisible to our eyes.

Some of the most spectacular 'fireworks' images were taken in infrared light. Infrared radiation is similar to visible light, but the wavelengths are longer. It is familiar to us as heat radiation. We cannot see it, but we can feel its action on our skin. Infrared radiation from the Sun can illuminate solar system objects, just as visible light does, especially at wavelengths shorter than about five microns, but warm objects also produce infrared radiation of their own. (One micron is 1000 nm.) The hotter an object, the more infrared radiation it emits overall, and the shorter the wavelength at which it radiates maximum energy. This means that infrared radiation can be used to measure temperatures.

The energy released in the explosions following the impacts raised the local temperature enormously creating bursts of infrared emission. Infrared observations can be made from ground-based observatories, preferably on high mountain sites, but most ultraviolet light, which has wavelengths shorter than visible light, does not penetrate our atmosphere at all. Only spacecraft, most notably the Hubble Space Telescope, could take ultraviolet images or spectra.

Most images made by professional astronomers today are recorded with electronic detectors rather than with photographic materials (though SL9 was discovered on a real traditional photograph). Charge-coupled devices (CCDs) and infrared detector arrays consist of a grid of individual small detection points, called pixels (short for picture elements). That is why astronomical images can look as if they are made up of rows and rows of little squares. The brightness of each pixel is stored as a number on a computer, which can reconstruct the numbers into an image when needed. In the reconstructed images, different brightness levels are often represented by different colors in a fairly arbitrary scheme. There are many of these 'pseudocolor' images in this book. They may have a blue or orange tint, for example, that is attractive but has no basis in reality.

Other color images combine information from several wavelengths so that the colors in the image give information about the wavelength of the radiation, just as in the case of the color images produced by our eyes.

In some of these cases, the wavelengths and their color representations are chosen to produce a close approximation to what we would see with our eyes.

The new electronic detectors are far more sensitive than old photographic film and have revolutionized observational astronomy in little more than a decade. Had SL9 crashed into Jupiter a few years earlier, we would not have been able to bring you the magnificent selection of images presented in this book.

2

A Comet Like No Other

Carolyn S. Shoemaker & Eugene M. Shoemaker

*'I don't know what this is, but it looks like a squashed comet.'
What Carolyn Shoemaker saw in her stereo-microscope was
certainly unlike any comet she had seen before. Soon the whole
world would know that its train of fragments was destined to
plunge into the planet Jupiter in July 1994.*

The discovery of Periodic Comet Shoemaker–Levy 9 was one of the fruitful outcomes of an intensive survey of the sky we started in 1983 – the Palomar Asteroid and Comet Survey. The goal of our survey was to determine the nature and numbers of objects that crash into the planets and satellites and form craters. Over the course of the 12-year survey, concluded in December 1994, our observing team took about 20 000 photographs of the sky with the Palomar 46-cm Schmidt Telescope. Henry E. Holt joined our team and participated part of the time from 1987 onwards. David H. Levy joined and participated in about half the observing runs, beginning in 1989. In the later years we were sometimes assisted by students from Northern Arizona University and elsewhere. All of the observers except Gene Shoemaker participated as volunteers. Generally we observed 7 nights a month, about 11 months per year. In the course of the survey we discovered more than 2000 asteroids, including 45 Earth-approaching asteroids, about 70 Mars-crossing asteroids, 150 asteroids with highly inclined orbits, 47 Trojan asteroids (members of distant swarms of asteroids preceding and following Jupiter) and a total of 33 comets. The most exciting

discovery was the thirtieth comet, which was also the ninth periodic comet found by the team of Carolyn and Gene Shoemaker and David Levy.

Discovery

The discovery of Comet P/Shoemaker–Levy 9 (SL9) on 25 March 1993 depended on a confluence of unusual factors, including a cloudy sky and a set of light-contaminated photographic films. The March 1993 observing run was the third in a row where bad weather prevented us from using much of our allocated time on the 46-cm Schmidt telescope at Palomar Observatory in San Diego County, California. The 46-cm Schmidt is a wide-field photographic telescope. Each film covers a circular field just under 9° across, equivalent to about 60 square degrees of the sky. (The entire sky is about 40 000 square degrees.) In the later years of the survey, we used a fine-grained emulsion, Kodak Technical Pan 4415 film. To increase its speed we hypersensitized the film by baking it for six hours at 65 °C in a dilute mixture of hydrogen and nitrogen gas.

During the first two cloudy months of 1993, we had accumulated a large supply of hypersensitized film. We

David Levy (left), Carolyn Shoemaker and Gene Shoemaker celebrate their discovery at the Palomar 46-cm Schmidt telescope. (Alan Levenson)

finally got the chance to use it on 22 March, the first night of our March run and the first completely clear night we had seen in a long time. On this run we were joined by Philippe Bendjoya, an astronomer from France. We began observing in a hopeful spirit, but all that was dashed as soon as Gene developed the first films. They were completely black. Sometime in the previous weeks, it seemed, someone had opened the sealed box in which we store the hypersensitized film and had exposed the film to light. And we had no fresh film ready.

The films were nested in a stack in the bottom of the box, and Gene quickly developed a couple of films from near the center of the stack. Although their edges were light struck, the center sections were usable. We resumed observing and got through the first night. The following evening, the 23rd, with freshly hypersensitized film, we began again. Things went smoothly for the first four hours, until the arrival of cirrus clouds from an approaching storm forced us to stop once more.

We stood outside the dome, gazing forlornly at the gathering clouds. David didn't think the sky was all that bad, but Gene tempered his enthusiasm. 'We spend almost four dollars each time we put film into the telescope,' he said. 'We simply can't afford to waste film on a bad night like this one has become. We have to wait till it clears.'

We were about ready to go back inside and wait when David thought of the light-struck films. 'Don't we have a few of those left from our first night?' David asked, aware that we had all but abandoned those films anyway. We looked back at the sky, which seemed to be getting a little better. 'Let's get to work!' Gene decided, and one of the partially light-struck films was quickly loaded into the telescope. The telescope was pointed, but David had difficulty finding the star on which he would guide for the eight-minute exposure. Jupiter was close by, and its glare swamped the field of the guiding eyepiece. Exposures of three fields of sky were taken, but as the clouds thickened, we had to stop again.

After two exposures have been made of a particular area of sky, Carolyn scans them using a special stereo-microscope. Just as in any stereo-photographic pair, an object that is in a different position in the two photographs will stand out clearly. When the films are properly aligned, a moving object appears to rise above the background of stars. The process is the same as looking at a pair of stereo-photographs of a landscape; when viewed in stereo, the nearby trees force themselves dramatically in front of the distant mountains. It takes Carolyn about 20 minutes, on average, to scan a stereo pair of films, but before she can scan she needs two films of the same field, preferably exposed about 45 minutes apart.

Getting that second film for those last three fields was a challenge. David and Gene went outdoors repeatedly to check the sky and after about an hour a small break in the clouds seemed to be moving in the direction of

The discovery images of SL9. When the pair of images is viewed stereoscopically, the comet appears to 'float' above the stars due to its motion relative to the stars in the one-and-three-quarter-hour interval between frames. (Gene and Carolyn Shoemaker)

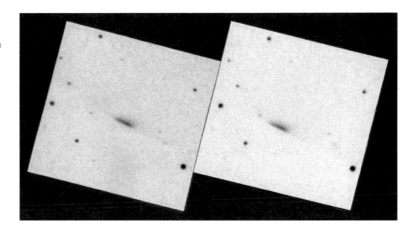

Jupiter. When it arrived they were ready. They obtained the three eight-minute follow-up exposures before the hole in the clouds vanished. Except for two sets of films that next night and a slight break near the end of the run, our observing was essentially wiped out for the rest of the scheduled nights.

By 4 p.m. on the afternoon of 25 March it was windy and cold. Having scanned the films from the first night, Carolyn was down to scanning the short set we had taken two evenings earlier. Gene and David returned from a trip to the hypering oven, which sits in a different dome. The thick clouds, fog, wind, and snow were getting depressingly familiar. 'We're through for this night,' David volunteered. 'What!' Gene said. 'Where is the Levy enthusiasm?' Carolyn was scanning the field with Jupiter on it. Then she sighted something near the center of the field, about 4° away from the planet. 'I don't know what this is, but it looks like a squashed comet,' she remarked.

The others crowded around to see for themselves. Totally bemused, they stared through the eyepieces. Hovering above the stars was something that looked like a comet, but instead of the round fuzzy coma that comets should have, this object looked like a bar of cometary light. We puzzled over this for some time. It couldn't be a ghost image of Jupiter, which appeared as a huge black blotch on the film. The ghost image was on the other side of the film where it belonged. Some other familiar optical artifacts such as diffraction spikes, were in their usual position. This bar-shaped object was completely different.

Confident that it was a comet of unusual form, Carolyn plotted the approximate position of the strange object; a description, together with the approximate celestial coordinates, were promptly sent by electronic mail to Brian Marsden, Director of the Minor Planet Center in Cambridge, Massachusetts. Then we went off to dinner, discussing among ourselves what could have produced the bar-like form. The comet was close to Jupiter in the sky and its motion mimicked that of Jupiter, suggesting that it might also be fairly close to Jupiter in space. Could it have passed so close to Jupiter that it had been broken up by the tidal forces of the jovian gravity field? It seemed a reasonable explanation, but the odds seemed very low that a comet had recently passed sufficiently close to Jupiter.

Confirmation

Returning to the 46-cm dome after dinner, we faced the next crisis: how do we go about confirming this strange comet? Our sky was supposed to remain cloudy for days and did. David had thought of our friend James Scotti, who was observing that very night on the Spacewatch Telescope at Kitt Peak in Arizona. We hoped that the eastward-advancing storm had not cut him off yet. We called Scotti that evening, and found that he was still in the clear. When Gene described the object to him, he doubted that we had a real comet. Its motion so closely matched Jupiter's, that he thought it might be an artifact due to some unknown reflection in the telescope of the bright light from Jupiter. However, he agreed to take some images with the CCD camera on the Spacewatch telescope.

In the meantime, Gene, David, and Philippe Bendjoya went to the dome of the 1.2-m Schmidt telescope to obtain better measurements of the celestial coordinates of the comet from our two available images. They were assisted by Jean Mueller, the principal observer on the 1.2-m Schmidt.

A CCD image of SL9 taken with the 0.91-m Spacewatch telescope at Kitt Peak on 30 March 1993 at by Jim Scotti. The multiple nuclei making up the central 'bar' can just be seen and prominent dust trains extend from either end of the bar. Dust, blown back by the pressure of sunlight, extends toward the upper right from the individual nuclei and the dust trains. (J. Scotti)

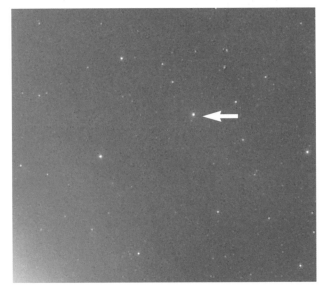

One of the SL9 discovery plates in its entirety (top). The comet is not visible on this scale but its location is marked by an arrow in the enlargement of the part marked by a box (bottom). Brightening around the edge of the plate is due to the accidental light leak that almost ruined this batch of plates, and Jupiter is the bright spot on the left. (Gene and Carolyn Shoemaker)

When they returned two hours later, they called Scotti again. 'Have you got yourselves a comet!' he said, sounding as though he had just seen the comet of his life. In fact, he had. From the images he had obtained with his larger telescope, he described its appearance, the multiple nuclei with individual tails and the wings, just as the fourth movement of Beethoven's first symphony was

wafting through our dome. The exultant music seemed to match Jim's words perfectly. After we ended the conversation, David renamed Beethoven's work the Comet Symphony, and after all these months, its notes are still a match of the wildest comet we have ever gazed upon.

Announcement

The announcement of the comet's discovery by Brian Marsden went out over the e-mail internet the next day as International Astronomical Union Circular 5725. Within days the comet was being followed by observers around the world. Some of the most exquisite early images were obtained by Wieslaw Wisniewski with the 2.3-m telescope at Kitt Peak, Arizona on 26 March and by David Jewitt and Jane Luu with the University of Hawaii's 2.2-m telescope on Mauna Kea, Hawaii, on 27 March. Jewitt and Luu described the comet as resembling a string of pearls, an appelation that stuck.

One of the first detailed images of the comet, taken on 28 March 1993 by the late Wieslaw Wisniewski with the University of Arizona's 2.3-m telescope at Kitt Peak. The total exposure time was 5 minutes and at least 12 individual nuclei can be seen. The field of view is 1.3 arcminutes across. (W. Wisniewski)

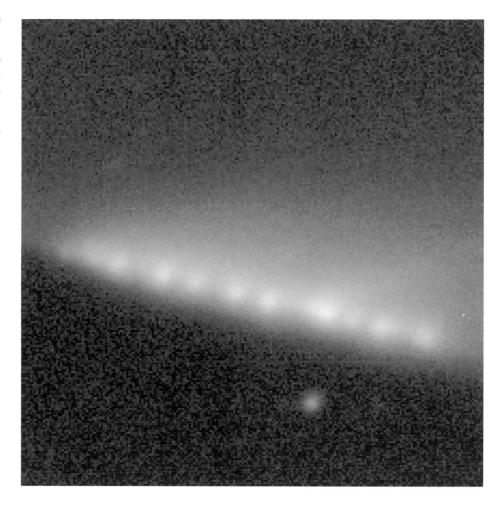

Meanwhile, other astronomers began searching films and photographic plates taken previously. In early April, Marsden announced positions of the comet measured from photographs by the Japanese observers Kin Endate and Satoshi Otomo taken as early as 15 and 17 March. Positions were also reported from films taken by Eleanor Helin and her colleagues with the 46-cm telescope, just prior to our own observing run on the telescope. A somewhat delayed though apparently independent discovery was made by Orlando Naranjo on plates taken with a 1-m Schmidt telescope in Venezuela, just hours before the discovery announcement went out on March 26. Perhaps the most regrettable missed opportunity for discovery was that of Gonzalo Tancredi of Uruguay and Mats Lindgren of Sweden, who had taken plates containing the comet's image on the 1-m Schmidt telescope of the European Southern Observatory at La Silla, Chile, on 19, 20, and 24 March. They were engaged in a deliberate search for comets near Jupiter, but SL9 was so strange in appearance that they did not recognize it. The occurrence of comets both near Jupiter in space and captured in temporary orbit about Jupiter was the subject of Tancredi's PhD thesis, then shortly to be completed at Uppsala University, Sweden.

Would the comet have been found if we had not taken those exposures with light-struck film? It is impossible to say with certainty. Playing 'what if' games with history is never a testable intellectual exercise. The odds are that, if the comet had not been reported beforehand by someone else, we would have found it ourselves on our next observing run. Had the weather been fine, very probably we would have found the comet in February. Indeed, bright as it was, the comet could readily have been found four months earlier than it was, had anyone bothered to look in the right place.

Jupiter's family of comets

Comets are remnants of solid material left over from the process of planet formation and come from the outer regions of the solar system. Because they contain ices of volatile compounds, chiefly of water, comets develop an extended atmosphere or halo of gas and dust when they approach the Sun. It is this atmosphere, called the coma, that distinguishes a comet from small inactive bodies orbiting the Sun that are referred to as asteroids or minor planets. Most discovered comets have arrived near the Sun from regions beyond the major planets. They follow very elongated orbits that take them back to these distant regions; they are observed in the inner part of the solar system only once. Some known comets, about 150 in all, have been captured by the giant planet Jupiter into short-period orbits through a succession of gravitational perturbations. This allows them to be observed on frequent repeated passages near the Sun. The average period of revolution about the Sun of these so-called Jupiter-family comets is about eight years.

Occasionally, after a long succession of encounters with Jupiter, a Jupiter-family comet can be captured into a temporary orbit as a satellite of Jupiter. Usually a comet makes only one or a few revolutions about Jupiter as a satellite before it escapes into independent orbit about the Sun and becomes, once again, a Jupiter-family comet. Comets P/Gehrels 3 and P/Helin–Roman–Crockett, for example, were discovered shortly after their escape from very short-lived temporary orbits about Jupiter. We know this, because their trajectories in space can be traced backward in time by mathematical methods of celestial mechanics. From various studies of the way the orbits of Jupiter-family comets evolve, and from available observations of comet nuclei, we estimate that, on average, at least one comet with a diameter greater than 1 km is loosely held captive as a satellite of Jupiter at any one time. Generally it is too faint and inactive to be detected by the ordinary methods used to search for comets. SL9 is the only comet ever discovered in a moderately long-lived orbit around a planet.

On 22 May 1993, Brian Marsden wrote in IAU Circular 5800 that SL9 would probably would strike Jupiter on its next close approach to the planet in July 1994. This announcement galvanized the astronomical community. By year end it was clear that collision was virtually certain.

3

The Path To Destruction

Brian G. Marsden

As position measurements started to accumulate from observers round the world, the catastrophic fate awaiting Comet P/Shoemaker–Levy 9 gradually became apparent. Orbit calculations revealed that it had been in Jupiter's power for decades. At too close an encounter in 1992 it was shredded apart and launched on its path to total destruction.

At its discovery in late March 1993 the comet that would be known as P/Shoemaker–Levy 9 was in several fragments. Only four degrees away from Jupiter, it followed more or less the same track across the sky as the planet. This strongly suggested that a close encounter between the two bodies had shattered the comet. But when and how had the comet come close enough to Jupiter to precipitate this catastrophe, and what was its future course?

To calculate the orbit of a comet with any degree of precision, we need a series of position measurements. The accuracy of the calculations improves as more observations are taken into account, and the interval of time they cover increases. It seemed likely we should have to wait for some weeks after discovery before there would be enough data to start building a definite picture of what had happened to this comet, and where it was now heading. As often happens, once the discovery had been announced, it turned out that there were earlier observations that had not been reported, in this case mainly because the images did not resemble a normal comet.

Pre-discovery images

The fragmented comet first made its presence known on Earth on 15 March 1993 by imprinting its image on a film exposed by the Japanese amateur astronomer Kin Endate with the 25-cm Schmidt telescope at his private observatory in Kitami. In collaboration with Kazuro Watanabe, Endate had been carrying out observations of asteroids since 1987. There were images of four new asteroids on the film, but the comet was not noticed at the time.

The comet and one of the new asteroids were also recorded on a pair of films taken by another Japanese amateur, Satoshi Otomo, in Kiyosato on 17 March. Otomo noticed the fuzzy image, but he thought it was a galaxy and did not check to see whether it was moving – until after he learned of the comet's existence.

A team from Uppsala Observatory in Sweden had undertaken a search of the region around Jupiter specifically for comets that might be in its vicinity at any time. This survey within a few degrees of Jupiter on plates taken with the 1-m Schmidt telescope at the

The first known image of SL9, unknowingly photographed by Kin Endate with a 25-cm Schmidt camera from Kitami, Japan, on 15 March 1993, nine days before discovery. There are two exposures on the photograph, separated by three minutes. Comets and asteroids reveal themselves by their motion between exposures. The comet is the faint bar in the center of the frame. The offset between the two comet images is different from that between the image pairs of the background stars, showing the comet's motion. (K. Endate & K. Watanabe)

European Southern Observatory had been initiated in 1992. The first plates had failed to show any comets, but the searchers decided to try again in 1993. A student from Uppsala, Mats Lindgren, was for the second year in succession despatched to Chile to examine the plates as soon as they had been taken. On one of the plates taken on 19 March Lindgren did see the fuzzy image of the comet, but he did not recognize it for what it was – something he later ascribed to his lack of experience. His action at the time was only to ensure that more plates of the same region would be taken on 20 and 24 March.

Eleanor Helin also saw the comet's image on 19 March on a film taken with the 46-cm Schmidt telescope at Palomar Observatory some six hours after the ESO–Uppsala plate. This was during the course of her survey for near-Earth asteroids. She marked it but, like Otomo and Lindgren, did not pursue the matter. She was preoccupied with her application for financial support to continue her NEO observations, which had a deadline looming later that day. So she ensured that a follow-up film of the region would be taken on 21 March. After completion of their observations on 22 March, Helin and

her assistants left Palomar. The Shoemaker team, engaged in a similar observing program, replaced them.

Recognition at last

When, on the afternoon of 25 March, Carolyn Shoemaker saw the strange image on a pair of films taken by her husband Gene Shoemaker, Tucson amateur astronomer David Levy and visiting French astronomer Philippe Bendjoya some 36 hours earlier, action was finally taken. They informed me at the Central Bureau for Astronomical Telegrams, giving a rough position and a general indication of the object's motion. They described the curious appearance of the image as 'a dense, linear bar very close to 1 arcminute long' but acknowledged 'one disconcerting aspect', namely, that 'Jupiter is in the same field, though many degrees away, and the motion of our suspect is rather close to that of Jupiter'. 'The image is certainly not a ghost image of Jupiter', they stated confidently, but 'the long axis of the bar is aligned somewhat toward the direction of Jupiter', so 'it is possible that we could have been fooled'. Confirmation of the comet that night by James Scotti using the

Spacewatch Telescope in Arizona was spectacular and clearly showed that the 'linear bar' consisted of at least five, and perhaps as many as 11, individual cometary nuclei.

Gene Shoemaker felt that the splitting had happened rather recently. I suspected it must have taken at least six months for the pieces to separate, and I presumed that the splitting had also caused the comet to brighten. Since I knew that the Uppsala–ESO survey in 1992 had not reported the comet, this suggested an upper limit of rather less than twelve months since fragmentation. The next day I announced the discovery of Comet Shoemaker-Levy (1993e) in International Astronomical Union Circular No. 5725, along with precise measurements of its position made on 24 and 26 March. At that time I computed, but did not publish, my first guess at the time of splitting, a compromise with an encounter date in June 1992. Using additional measurements made the next night I suggested encounter on 28 July 1992, although it seemed likely that my distance of closest approach, 0.04 AU or 80 Jupiter radii (R_J), was too great for Jupiter's gravity to have broken up the comet.

Uncertain though the circumstances of the encounter were, there was some value in showing that an appropriate encounter scenario was viable. At this stage, we were taking for the position of the comet the centre of a string of bright points extending over an arcminute. We just did not know the comet's location in space, so it was impossible to work out how much of its acceleration was due to the gravitational influence of Jupiter and how much was due to the Sun. Indeed, the observations available at this time were fully consistent with a parabolic orbit around the Sun that took the comet nowhere near Jupiter!

On 31 March we learned of a true independent discovery of the comet by Orlando Naranjo using the 1-m Schmidt at Mérida, Venezuela. A collaborator in the Uppsala–ESO survey, but unaware of the Shoemaker–Levy announcement, he had found and measured images of the comet on 26, 27 and 28 March. In late March amateur astronomers visually estimated the comet's total magnitude as about 13, too faint to have made the prospect of an earlier visual discovery reasonable.

A satellite of Jupiter

By 3 April, measurements were available up to 1 April and from most of the exposures showing the comet back

to 15 March. A parabolic orbit about the Sun no longer fitted the accumulated observations, but the eccentricity of any elliptical orbit had considerable uncertainty. I came up with a possible solution that had the comet pass only 0.007 AU (14 R_J) from Jupiter on 16 May 1992. The encounter distance was still too large, but the result was interesting because it gave the first hint that the comet was actually orbiting Jupiter.

Comets had been known to be captured into temporary orbits about Jupiter before. I had noted that P/Oterma made a partial orbit about Jupiter in the early 1960s, something that had struck the fancy of Seth Nicholson, the Mount Wilson astronomer who discovered four permanent jovian satellites between 1914 and 1951. P/Gehrels 3 was in a loose orbit around the planet in the early 1970s, coming as close as 0.0014 AU (2.8 R_J) in August 1970 and to 0.04 AU in March 1973, after which it escaped from Jupiter and was discovered in 1975. P/Helin-Roman-Crockett, discovered in 1979, made several orbits of Jupiter between 1976 and 1983, coming as close as 0.02 AU.

At discovery, P/Shoemaker–Levy 9 (as it came to be called), or 'SL9', was apparently close to apojove (its furthest distance from Jupiter), some 0.31 AU from the planet. Presumably there would be another Jupiter encounter in 1994, but how close would the encounter be then? Using a good series of measurements by Japanese amateur astronomers, Syuichi Nakano was finding in mid-April that the computations inexorably gave a Jupiter encounter in 1992, without any need to make assumptions. But his encounter date of January 1992 had to be wrong, because it preceded the Uppsala–ESO observations that did not show the comet. With observations extending to 25 April, Nakano got an approach within only 0.004 AU (8 R_J) of Jupiter on 11 July 1992 – and the possibility of an even closer approach in July 1994. An encounter at 0.004 AU was still not within the Roche limit, the maximum distance from the planet that would effectively guarantee some break-up.

On 7 May I wrote to Nakano, 'I do think we are now starting to converge on the truth. There are a few more observations now (extending to 30 April), and I am also getting Jupiter encounters, to about 0.001 AU (the Roche limit) in good cases, in late June or the first half of July 1992... Maybe the comet will hit Jupiter in 1994, but the chance would seem small. We need more observations!' Almost all the observations made in April and May were by amateur astronomers in Austria, Italy and Japan. Given the fact that some of these observations were photographic and that it was difficult to pick out individual nuclei even in the CCD images, the convention of using measurements only of the centre of the train of nuclei was continued. Both Nakano and I progressively refined the orbit computations, although the fact that different observers were calling different points in the train of fragments the 'center' made a definite conclusion difficult.

In mid-May Andrea Carusi in Italy was trying to follow the orbital motion of SL9 backwards through the 1992 encounter. What impressed him was the length of time the comet seemed to remain a satellite of Jupiter in comparison with other comets known from subsequent observations to have been satellites temporarily at an earlier time. There was uncertainty in the details, of course, but his conclusion has subsequently been justified.

The comet crash predicted

When the observations extended to 18 May I was almost completely convinced that a collision with Jupiter would occur and announced this possibility in an IAU Circular on 22 May. If the center of the train were to hit Jupiter, Carusi noted, the whole train of fragments would strike the planet over the interval 23–27 July 1994. Donald Yeomans and Paul Chodas, starting independent

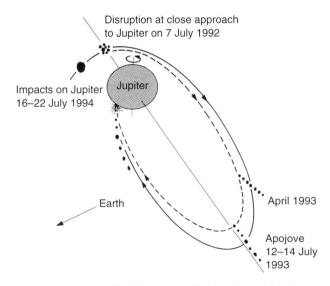

A schematic diagram of SL9's break-up and final orbit around Jupiter (not to scale). (Z. Sekanina, D. Yeomans & P. Chodas/P. Doherty)

computations on the comet's orbit at the Jet Propulsion Laboratory (JPL), at first estimated the probability of a hit as 64%, but after only a few days raised it to more than 99%.

Although amateur observations giving positions for the center of the train continued into July 1993, measurements of the individual nuclei were badly needed. Working with the 2.2-meter reflector at the Mauna Kea Observatory in Hawaii, David Jewitt and Jane Luu noted that the comet had as many as 21 fragments on their CCD image of 27 March. In September they provided relative measurements of the individual nuclei from that and three other images obtained in April, June and July. Harold Weaver produced relative measurements of 11 fragments from an image obtained using the Hubble Space Telescope on 1 July.

From these relative measures Zdenek Sekanina at JPL attempted the first analysis of the circumstances surrounding the splitting of the comet. He also offered predictions for the individual impacts in 1994, which, to everyone's frustration, would occur on the side of Jupiter facing away from Earth. The six or eight brightest fragments, he concluded, had separated from each other a couple of hours after the comet's passage only 0.0006 AU (90 000 km) from Jupiter on 7 July 1992. This meant they were each perhaps 2–4 km in diameter. These components were lined up rather precisely. Other fragments located off the alignment were the outcome of later separations. The relationship between the diameters of the fragments, their speed of separation, and time since perijove, now became a matter of debate. Jay Melosh and Jim Scotti thought that the original break-up was closer to perijove and the diameters more likely in the range 0.5–1 km, a conclusion also reached by Graeme Waddington in England. According to these two different scenarios, the original nucleus would have been either 8 km or 2 km across. The failure of the comet to show on the 1992 Uppsala–ESO plates did not resolve the discordance. It simply indicated that the nucleus was not larger than 15 km.

In any case, the problem remained that the computations were based ultimately on measurements of the 'train centre', a fictitious and variable point. In mid-November James Scotti, assisted by Travis Metcalfe, provided absolute measurements of nine of the fragments on up to nine nights between 30 March and 19 July. Orbits computed just from these observations narrowed down the predicted dates for the impacts of fragments designated E, G, H, K, L, Q, R, S and W to the period 18–23 July 1994. A tie-in with the relative positions from Mauna Kea indicated that fragments A–D would hit Jupiter during 17 and 18 July.

Just as James Scotti had been the last to record the comet before it disappeared into the glare of the Sun, in the evening sky on 21 July, so was he the first to pick it up when it reappeared in the morning sky on 9 December. Scotti's observations on 9 and 14 December indicated that the previously predicted impact dates were about 24 hours too late. Most exciting, Yeomans and Chodas calculated improved impact points and found that the impacts would occur only 5–10° behind the visible disk, with later impacts being closest to the limb. It would thus take only 12–24 minutes for the impact sites to rotate into view, giving the first hope that the later stages of the impact process might be directly visible from Earth. The *Galileo* spacecraft, as had been hoped, would have a direct view of the impacts. It also seemed that two of the 21 fragments noted by Jewitt and Luu, J and M, had vanished.

The final run-in

During January–July 1994, 18 different observatories (three of them operated by amateur astronomers), including the HST, made a total of 2658 observations of the 19 remaining nuclei. The orbits were further improved by Nakano, by myself and particularly by Yeomans and Chodas at JPL, who with the help of Sekanina produced and distributed detailed predictions of the geometrical circumstances of the impacts on Jupiter. Fragments G, P and Q broke up further.

In an intense campaign at ESO starting on 1 July, Richard West and his colleagues measured the positions of fragments using for reference as yet unpublished star positions in the catalogue from the recent *Hipparcos* satellite mission. These data are more consistent than the *Space Telescope Guide Star Catalog* used by the other observers. Others also made critical observations during these final days. On 19 July, David Jewitt measured fragment P and beyond after several of the earlier fragments had already crashed into the planet.

At 16:00 Universal Time on 16 July Chodas and Yeomans issued their final impact predictions before the collisions started, giving an estimated time for Fragment A's demise of 20:00 UT later that same day, with an uncertainty of about six minutes. Fragment A actually struck Jupiter around 20:11 on 16 July, and by 08:06 on

A painting by artist Don Davis of the view towards Jupiter as he imagined it to be from the nucleus of one of the fragments, three days before the impacts.

22 July the whole comet had gone. The final predictions had on average been accurate to about 7 minutes. The principal cause of the error was probably the inadequacy of the star positions in the *Guide Star Catalog.*

1992 and before

With the final orbits of the pieces known, computations following the comet's motion backwards in time to before the fatal 1992 encounter became more meaningful. Unfortunately, previous orbits derived from such calculations will always be rough approximations, because small uncertainties in the comet's position during the 1992 encounter make for large uncertainties in the degree of bending of the trajectory during the encounter, and hence in our projections of the orbit before encounter. Many investigators have made such computations and there is widespread agreement on the general outlines of what happened, working back through several perijove passages that were quite far from Jupiter to the close approaches which occurred in 1972 and 1970.

Although its history prior to 1970 is even more uncertain, indications are that SL9 was then already orbiting Jupiter. Lance Benner and William McKinnon of

Washington University, St Louis, Missouri calculated one possible scenario for the trajectory of the comet during this time. Formal calculations suggest that it may have been a satellite of Jupiter even before 1914. A statistical analysis of the problem by Yeomans and Chodas concludes that, when captured by Jupiter, the comet was in a nearly circular orbit, tilted only slightly to the main plane of the solar system, and its distance from the Sun was probably 4 AU (or possibly 6–8 AU). It was almost certainly captured before 1960 and most likely in the period 1920–1930. Because it was very weakly bound to Jupiter, its orbit was exceptionally chaotic. Nevertheless SL9 lasted longer as a jovian satellite than any other known comet.

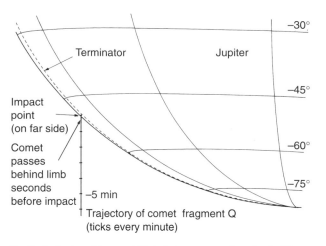

The final minutes of fragment Q's trajectory, as seen from Earth. All fragments followed almost identical trajectories. (P. Chodas/P. Doherty)

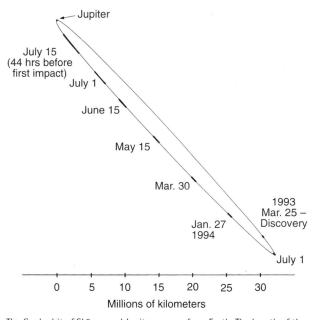

The final orbit of SL9 around Jupiter as seen from Earth. The length of the comet train is shown at various times by the heavy black bars. The comet stretched dramatically as impact approached. (P. Chodas/P. Doherty)

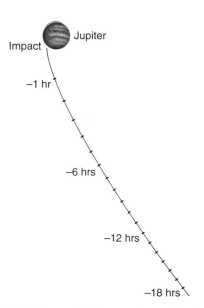

The final approach of fragment Q toward Jupiter, as seen from Earth. All fragments followed almost identical trajectories. (P. Chodas/P. Doherty)

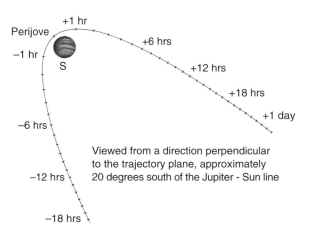

The circumstances of SL9's destructive encounter with Jupiter in 1992, calculated by Paul Chodas after the impacts in 1994.

4

The Giant Target

John R. Spencer

Jupiter is a planet totally different from Earth. Essentially a giant gas ball, it has no solid surface beneath the swirling colored clouds we see. With its powerful gravity and strong magnetic influence, it controls and interacts with a fascinating family of moons.

Jupiter, the solar system's largest planet, is the centerpiece and controller of a system of extraordinary and fascinating complexity. As Comet Shoemaker–Levy 9 bore down on this giant planet, astronomers shared the hope that they would learn more about what makes the Jupiter system tick by watching its response to the intruder.

Origin

Jupiter contains 70% of the material in the solar system, not including the Sun itself. We think it was born along with the Sun and the other planets in the solar nebula, a disk of hot gas and dust that surrounded the nascent Sun. As the nebula cooled, rock and metal grains began to condense. These grains collided and coalesced, and were drawn together by gravity, to form the seeds of the planets. In the inner solar system the assemblage of chunks of rock formed the terrestrial planets, including our own Earth.

At Jupiter's distance from the Sun, the accumulation continued until the gravity of the growing proto-planet was powerful enough to begin pulling in gas directly from the solar nebula. The mass of the added gas increased the gravitational pull and drew in yet more gas. This runaway process soon produced a world dominated by gas with essentially the same composition as the solar nebula. So Jupiter, like the solar nebula and the Sun, is mostly hydrogen. About 6% of its atoms (equivalent to 20% of its mass) are helium and there are small amounts of other elements.

Structure

Now let us take an imaginary journey outward from the center of Jupiter. We begin at the core of rock and iron that triggered the runaway accumulation of gas. Now it accounts for only about 4% of Jupiter's mass and is compressed under a pressure of up to 100 million Earth atmospheres by the weight of the overlying hydrogen. Above this core is the lower part of the hydrogen layer. At such high pressures, electrons flow freely from atom to atom, so that the hydrogen conducts electricity and behaves like a metal. Electrical currents flowing in this metallic hydrogen create Jupiter's powerful magnetic field.

About 70% of the way up to the cloud tops, where the pressure is a mere few million atmospheres (though we don't know the exact value), hydrogen nuclei pair up with electrons to form 'normal' hydrogen, which is a liquid at these pressures. Further out, the hydrogen is gaseous. There is no sharp transition between liquid and gas that we could call a surface. In the cooler upper regions where sunlight begins to penetrate, Jupiter's minor chemical constituents, which are locked in methane, ammonia, water, and a hundred other

Jupiter as seen by the *Voyager 1* spacecraft on its approach to the planet in 1979. The smallest details visible are about 600 km across. The subdued colours of this version of the image are more accurate than the oranges of the original *Voyager* press releases, though the many subtle shades seen through a telescope are not well represented even here. The Great Red Spot is prominent, and Io is visible on the right. (Charles Avis, JPL/NASA)

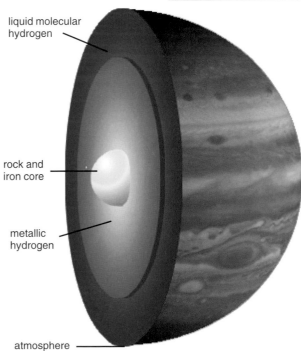

A schematic view of Jupiter's interior, showing the inner core of rock and metal, the metallic hydrogen layer, and the outer envelope of molecular hydrogen. The atmosphere is too thin to show on this scale. (J. Spencer)

compounds, begin to condense out to form multicolored clouds. Finally, high in the atmosphere, there are no more cloud-forming impurities. The cloudless upper atmosphere, still predominantly hydrogen, extends outward to merge imperceptibly with the vacuum of space.

We infer much of this picture from our knowledge of physics, from our ideas about the formation of the solar system, and from measurements of Jupiter's gravity and magnetic field. However, the uppermost layers, from pressure levels of a few bars outward, can be observed directly from Earth or from spacecraft. Since Jupiter has no solid surface, there is no clearly defined base above which to measure altitudes in the atmosphere. Instead, we commonly use the atmospheric pressure, which decreases with altitude, as a measure of altitude in the atmosphere. Pressure is measured in bars: one bar is roughly standard atmospheric pressure at sea level on Earth.

Atmosphere

The fantastically complex bands and swirls of Jupiter's clouds are the most striking features of its atmosphere.

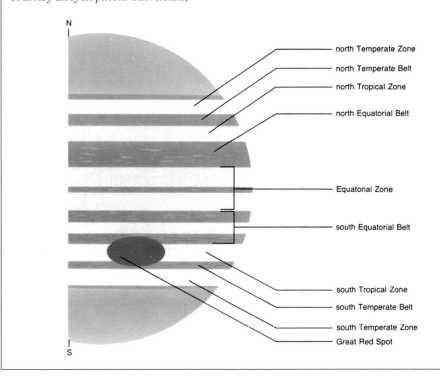

Early visual observers of Jupiter adopted names for the alternate light and dark cloud bands based on the climatic zones at the corresponding latitudes on Earth. Dark bands are called 'belts' and light bands 'zones'. Though belts sometimes vanish or change temporarily, the overall banded appearance of the planet has been much the same since it was first observed, and the nomenclature is still used. The main dark belts are easily visible through a small telescope. (Diagram from *The Cambridge Atlas of Astronomy*, courtesy Encyclopaedia Universalis)

Though visible directly with the human eye through even a small telescope, we could not see their full glory until the two *Voyager* spacecraft took close-up pictures in 1979. We are still very ignorant about the composition of the clouds, though it seems likely that the highest layers are crystallized ammonia cirrus and the lower clouds may consist of ammonium hydrosulfide, with water below that. We are even vaguer about what is responsible for the subtle brown and pinkish colors of the clouds (usually exaggerated to yellows and oranges in *Voyager* color reproductions) but sulfur is one possibility.

The clouds are in constant motion, driven by wind patterns quite different from their terrestrial equivalents. Jupiter's rapid rotation aligns the winds into alternating eastward and westward streams, with wind speeds reaching 400 km/hour. Superimposed on these alternating jets are numerous storms, symmetrical oval high-pressure systems and chaotic, twisted low-pressure regions. The most famous high-pressure system is the Great Red Spot, an extraordinarily long-lived storm that has been seen from Earth for 300 years and may be much older, but there are many other similar features of smaller size and differing color. The long life of the Great Red Spot and the other storm systems is all the more striking when compared with more chaotic regions on Jupiter, where cloud features a sizeable fraction of the dimensions of the Great Red Spot are created and torn apart in a matter of days. Despite the stability of the major spots, dramatic changes in Jupiter's appearance occur from year to year as bands of cloud form, spread, and dissipate at various latitudes.

The clouds extend upwards to pressure levels of about 0.5 bar. Above them the atmosphere is relatively clear, but hazes persist even here. We can study the high-altitude hazes from Earth by exploiting the fact that certain wavelengths of infrared light are absorbed very strongly by the methane that makes up about 1.6% (by mass) of Jupiter's atmosphere. If we look through filters

A schematic view of Jupiter's outer atmosphere. Thin hazes lie above cloud successively deeper cloud layers. The composition of these clouds is uncertain but they may be made of solid ammonia, brownish ammonium hydrosulphide, and water.
(J. Spencer)

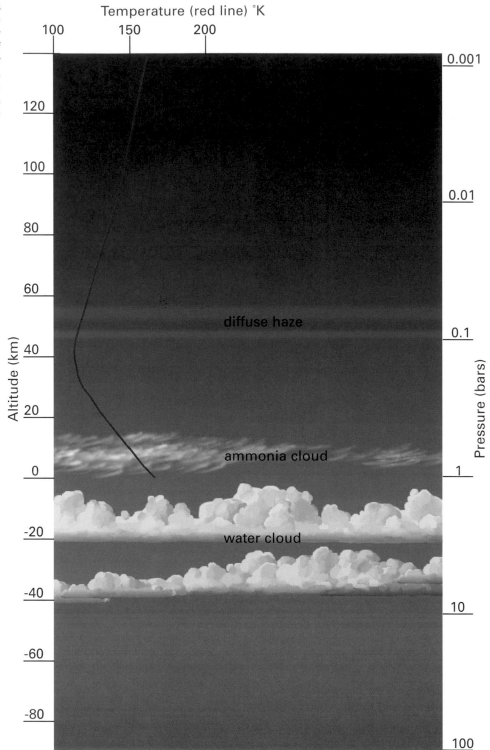

A *Voyager 2* close-up of Jupiter's clouds in the region of the Great Red Spot. (JPL/NASA)

that pass only those particular methane wavelengths, most of Jupiter looks almost black because sunlight falling on the atmosphere is absorbed by the methane molecules. However, if anything is high enough in the atmosphere to reflect this light back before it is absorbed, it will show up very brightly against the blackness of surrounding regions. When we make images of Jupiter at these methane wavelengths we see that the polar regions are bright, indicating high hazes there at an altitude of about 10 millibars. They may be connected in some way to Jupiter's aurorae. Also visible but less conspicuous are bright bands near the equator and over the Great Red Spot, revealing high hazes there also. The technique of imaging Jupiter at methane wavelengths proved very valuable for studying the comet impacts.

Infrared emission

Like Earth, Jupiter absorbs visible sunlight, which warms the planet and causes it to radiate heat at infrared wavelengths. In the case of Earth, the amount of energy absorbed in the form of visible sunlight is almost

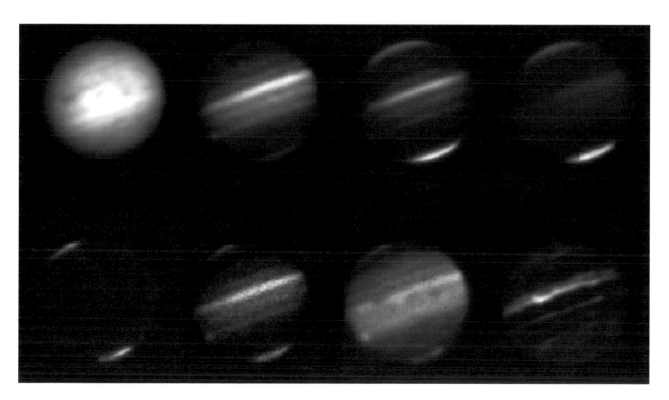

Jupiter at eight different near infrared 'colors', imaged by the NASA Infrared Telescope Facility at Mauna Kea with the NSFCAM array camera on 12 July 1994. Along the top row, (1.60, 2.04, 2.10, 2.27 microns) the effects of absorption by methane and hydrogen in Jupiter's atmosphere get progressively stronger and the images correspond to higher levels. In the far right image only the stratosphere is clearly visible. The bottom row (3.41, 3.80, 4.00, 5.05 microns) shows the polar aurorae in the first three pictures, with increasing amounts of reflection from clouds. In the last image, the bright areas are mainly Jupiter's own heat radiation escaping through less cloudy regions. (NASA IRTF Comet Science Team)

A high resolution ground-based image of Jupiter at a wavelength of 5 microns from the NASA Infrared Telescope Facility, Mauna Kea, in January 1994. Bright spots are due to heat radiation escaping from Jupiter's interior in regions where the clouds are thin. (John Rayner & John Spencer)

identical to the infrared energy that Earth returns to space: there is almost perfect balance. This is not so on Jupiter, however. Jupiter's heat radiation at longer infrared wavelengths is twice as bright as would be expected just from the sunlight that it absorbs: the planet must have a major internal source of energy to power the excess infrared radiation.

It seems likely that Jupiter is still cooling down from the high temperatures it reached when it first formed. The planet's slow contraction as it cools releases the extra energy. Jupiter's heat radiation can be seen most dramatically at a wavelength of five microns, where the glow of the warm, lower atmosphere silhouettes colder opaque clouds higher up and reveals intricate detail in the cloud patterns. At longer wavelengths the radiation originates from well above the clouds and Jupiter's appearance is less spectacular. Nevertheless, images and spectra of this heat radiation provide important details about the temperature and composition in the atmosphere, and the way that the planet rids itself of the enormous amounts of energy being generated in its interior.

Magnetosphere

Jupiter's domain extends far beyond the limits of its atmosphere to an array of rings and satellites, and an enormous and complex magnetic field. The magnetic field is generated by electrical currents in the metallic hydrogen in Jupiter's interior, in the same way that currents in Earth's iron core generate Earth's field, but at the cloud tops Jupiter's magnetic field is 20 times more powerful than at Earth's surface. The magnetic field extends far beyond the planet to fill a region known as the magnetosphere. Its ultimate boundaries are set by the impinging solar wind, the tenuous flow of atoms that streams constantly away from the Sun. Jupiter's magnetosphere extends 5 million kilometers in the upstream direction towards the Sun and has a tail stretching as far as the orbit of Saturn in the opposite, downstream direction. Because it is locked to the planet, Jupiter's magnetic field rotates every 10 hours, just as the planet itself does. This rotation has a powerful effect on the region around Jupiter.

Satellites and rings

The most prominent members of Jupiter's entourage are its four planet-sized moons, known collectively as the Galilean satellites. They are like a miniature solar system of remarkable diversity orbiting around Jupiter. The satellites probably formed at the same time as Jupiter from a disk of dust and gas surrounding the growing planet. The outer satellites contain more ice than the inner ones because ice could condense in greater abundance in the cooler outer regions of the disk. Jupiter's gravity has slowed the rotation of its satellites so each now keeps one face permanently turned towards Jupiter, just as Earth's Moon always keeps one face turned towards us. These satellites can be seen easily, even with small binoculars, as bright points of light close to Jupiter, but we only learned about their true personalities in 1979, thanks to the close-up images taken by the *Voyager* spacecraft.

Callisto, the outermost moon, is a cratered ball, half rock and half ice. It is a remarkably passive world, which has spent most of the history of the solar system accumulating on its surface impact craters due to the blows of incoming comets and asteroids. There they remain, since Callisto generates no internal volcanic or tectonic activity to erase them. No other object as large as Callisto shows such a density of craters.

The next satellite in is Ganymede, slightly larger and rockier than Callisto. Larger than Mercury, it is the largest moon in the solar system. Parts of its surface are also heavily cratered and probably date back to the earliest epochs of the solar system, but about half the surface has

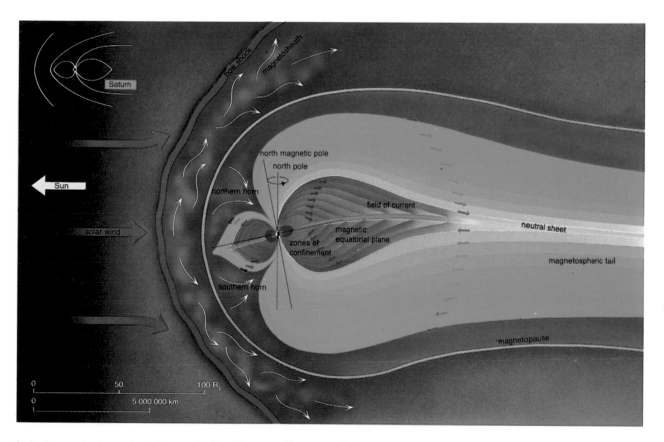

Jupiter's magnetosphere extends for several million kilometers. If it were visible from Earth, it would appear larger than the Sun or Moon. The bow shock is where the charged particles in the solar wind are slowed down and deflected around the jovian magnetic field. On the side facing away from the Sun, the magnetosphere is stretched out into an enormously long tail, which may stretch as far as the orbit of Saturn. (From *The Cambridge Atlas of Astronomy*, courtesy Encyclopaedia Universalis)

been extensively re-worked by a complex system of fault lines and volcanoes, though the volcanoes probably erupted water, not lava. Why Ganymede and Callisto are so different on the surface, yet so similar in size and composition, is one of the major unsolved problems of the jovian system.

Next comes Europa, a very peculiar world that is mostly rock on the inside but almost pure water ice on the outside. Its extremely smooth, crater-free surface indicates internal activity of a mysterious nature, The brilliant icy surface is covered in straight dark lines and a complex network of low ridges, some with curiously perfect geometrical shapes. The icy surface may cover a 100-kilometer-deep ocean of liquid water.

Io, the innermost Galilean moon, is the most remarkable satellite of all. Continual distortions of Io's shape by Jupiter's gravity heat much of the rocky interior to melting point, producing frantic volcanic activity at the surface. The two *Voyager* spacecraft spotted nine active volcanic eruptions during the 1979 flybys. Io's surface appears to be coated by white, pale yellow, and brown compounds of sulfur, spewed out by the volcanoes and probably overlying volcanic rock.

Io has a major influence on the whole Jupiter system, because the volatile compounds that cover its surface can escape into space, probably by way of Io's tenuous sulfur dioxide atmosphere. Atoms of sulfur, oxygen, sodium, and other elements leak from Io at a rate of about a tonne per second. They form a huge glowing cloud that is visible from Earth, especially in the orange-yellow light emitted by sodium atoms. Atoms in this cloud lose their electrons and become electrically charged. These charged atoms (ions), mostly sulfur and oxygen, are instantly caught in Jupiter's magnetic field, which, locked to the rotating planet, sweeps past Io at 70 km/s. Now also travelling at 70 km/s, the ions are spun out into a ring

The four Galilean satellites of Jupiter as seen by *Voyager 1*, at the correct relative sizes and roughly correct relative colors and brightnesses. They are Io (upper left), Europa, (upper right), Ganymede (lower left), and Callisto (lower right). (Charles Avis, JPL/NASA)

(called the plasma torus), which extends all the way around Jupiter at the distance of Io's orbit. It glows brightly enough to be visible from Earth. The ions in the torus then bombard Io's surface and atmosphere to create new ions and maintain the process.

The effluent from Io's volcanoes spreads beyond the visible torus to fill Jupiter's magnetosphere with high-speed sulfur and oxygen ions. These bombard the other satellites as well, staining and eroding their surfaces. The electrons that were stripped from the atoms when they were ionized near Io are also caught in the magnetic field and join the dance, sending out powerful bursts of radio energy as they spiral around Jupiter. As a result, Jupiter is one of the strongest sources of radio waves in the sky.

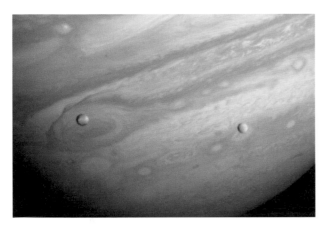

Sulfurous Io (brown) and icy Europa (white) orbit across the face of Jupiter's southern hemisphere in this spectacular 1979 *Voyager 1* image, recently improved by reprocessing of the original data. Io hangs 350 000 km above the Great Red Spot. The impacts occurred 15 years after this picture was taken at the latitude of the southernmost pale band visible here, just south of the band of alternating cyclones and anticyclones that can be seen behind Europa. (Charles Avis, JPL/NASA)

Some of the electrons and ions in the magnetosphere, still caught by the magnetic field, are flung into Jupiter's atmosphere at its poles. They make the atmosphere glow, forming aurorae like those that light up Earth's polar skies. However, the bombardment that causes Earth's polar aurorae comes from the Sun, while at Jupiter the ultimate source of much of the bombarding material is Io's volcanoes. Various forms of hydrogen in the aurorae glow with light of different wavelengths. The aurorae are particularly prominent in the ultraviolet (at a wavelength of 160 nanometres), where they are best seen by the Hubble Space Telescope, and in the infrared (at a wavelength of 3.4 microns), where they are particularly easy to study from Earth with infrared cameras.

Between Jupiter and Io, four small inner satellites orbit the planet, zipping around Jupiter with periods between 7 and 16 hours. With the exception of Amalthea, the largest at 270 km long, they were unknown till the

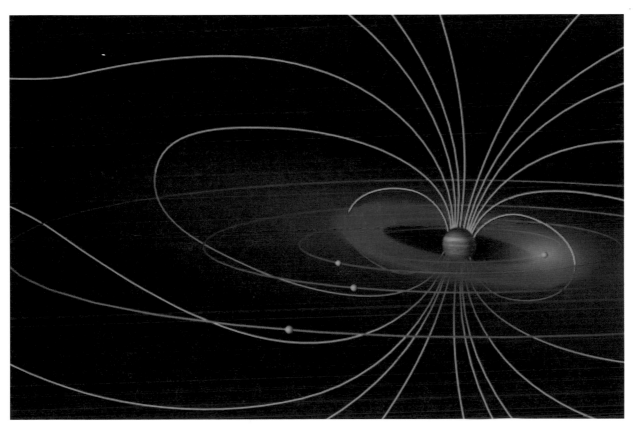

A schematic view of Jupiter's inner magnetosphere, to scale. The orbits of the four Galilean satellites are shown in orange. The volcanoes on Io, the innermost Galilean satellite, produce sodium which escapes to form a glowing yellow cloud around Io. Ions of sulfur and oxygen from the eruptions on Io are spun into a ring or torus (red) by Jupiter's magnetic field (field lines in blue). Jupiter's magnetic poles are tilted ten degrees away from the spin axis, so the field and torus wobble as Jupiter rotates. Ions are flung out from the torus by centrifugal force, distorting the magnetic field as they go, and forming a tilted sheet of ions (faint red) that bombard the other Galilean satellites. The magnetic field, which in reality fills all the space near Jupiter, is shown here in only two dimensions, for clarity. (J. Spencer)

Voyager spacecraft flew past Jupiter in 1979, but all are visible from Earth with the help of modern infrared cameras.

Jupiter's faint ring was also a *Voyager* discovery but can be seen from Earth, with difficulty. It is probably composed of fine dust ejected from the inner satellites and unseen smaller bodies by meteorite impacts. The dust particles spiral in towards Jupiter under the influence of the ions in the jovian magnetic field. In its inner regions,

the hail of electrons from the magnetosphere gives the dust an electrical charge and lifts it out of Jupiter's equatorial plane to form a dusty halo. The ring is so tenuous that 99.997% of the light passing through it is unaffected by its presence.

A world of complexity

So the world that captured Comet Shoemaker–Levy 9, tore it apart, and finally engulfed it in July 1994, is a place of dazzling complexity and constant activity. From Earth we can watch much of this activity in the form of storms boiling in Jupiter's clouds, heat radiating from its interior, volcanoes erupting on Io, the glow of the sodium cloud and sulfur ion torus that surround Io, the radio hiss and crackle of electrons caught in Jupiter's magnetic field, and the glow of the aurorae at Jupiter's poles. Jupiter is such a busy place that, before the impacts, we weren't sure whether we would be able to sort out the effects of the comet collisions from all the normal activity there but, on the plus side, Jupiter gave us a wide variety of phenomena to monitor. With so many things to look at, the chances that at least one of those phenomena would be affected by the collisions seemed good. In the everyday world, we poke and prod things around us to learn more about them. We hoped that the prodding of Jupiter by SL9 would tell us more about the nature of this extraordinary planet.

A pre-impact image of Jupiter taken with the NSFCAM array camera on the NASA Infrared Telescope Facility at Mauna Kea through a 2.26-micron filter to enhance the detection of the ring and the faint inner satellites. Amalthea is on the left of Jupiter and Metis on the right. (John Rayner/NASA IRTF Comet Science Team)

5

Bangs or Whimpers?

Kevin Zahnle

Faced with the prospect of the forthcoming comet crash, theorists had a little over a year to meet the challenge of predicting what observers might see. Simulating such a complex phenomenon proved very difficult even with the help of big computers. Even so, the theorists had a pretty good idea what the sequence of events would be. It was the spectacular consequences – infrared fireworks and huge dark scars – that came as a surprise to everyone.

The unique and wonderful 'string of pearls' comet was discovered in March 1993. Its fate was announced to an astonished planetary science community two months later: P/Shoemaker–Levy 9 was going to collide with Jupiter in July 1994.

At first the news hardly seemed credible. This was unprecedented in human history: a cosmic collision between two known bodies. The air of unreality was compounded by giddy early estimates that the comet's diameter was 10 or even 15 km. At that size the impacts would rival the impact that wiped out the dinosaurs here on Earth 65 million years ago. Such enormous events are obviously quite rare. Even on Jupiter, one would not expect something this large to hit more often than once in 100 000 years. Could we really be so lucky? In a word, 'No'. Later and more sober size estimates were in the range of a kilometer or two. But we were lucky enough: this was perhaps a once-per-thousand year chance to test our ideas about large impacts with hard observational data.

Once the initial astonishment had subsided, heads swam with questions, some specific to this comet, others we thought we'd never see answers to in our lifetimes. How big was this comet in reality? What was it made of? How did it get broken into so many pieces? What really happens in a big impact? What should we look for? Would there be anything to see? Can a scientist get a shoe advertising contract? (To avoid any unnecessary suspense on this last question, we came close. We were contacted by Reebok during the height of the feeding frenzy. But they lost interest two weeks later. I'll always regret this.)

This chapter addresses the problems theorists faced in making predictions, and why they seemed so wrong, when in truth they were actually pretty good.

Analogs on Earth and Venus

Jupiter is an airball. It has no surface. All impacts, however large, result in airbursts, i.e. explosions in the air. Airbursts on a much smaller scale have been observed to occur (or caused to occur) on Earth, and airbursts on a larger scale have been inferred to occur on Venus. By chance or cosmic design, my colleagues and I had recently completed studies of airbursts on Venus and Earth in the year before the discovery of SL9. We were ideally suited to issue early predictions, and ideally situated to be proved wrong by subsequent events. Serendipity was going to put our model to the test.

By far the best terrestrial analog to what was going to happen on Jupiter took place over central Siberia on the

The devastation caused in the Siberian Tunguska region when a meteoroid exploded 5–10 km above the ground on 30 June 1908. (TASS, courtesy Encyclopaedia Universalis)

morning of 30 June 1908. A meteoroid about 60 meters across exploded some 5 or 10 km above the central Siberian forest with the force of a 15-megaton thermonuclear device (roughly equivalent to the largest weapon). The blast wave from the explosion flattened trees over 2000 square km, and it remained strong enough to break windows 400 km away. Heat rays from the explosion caused a man 60 km away to pull his shirt off for fear it would catch fire. People who had no idea what was happening thought the world was about to end. Because the meteoroid itself was never found, the famous Tunguska explosion has been a magnet for weird science, but my colleagues Christopher Chyba, Paul Thomas and I developed a simple model to show that a Tunguska-like atmospheric explosion is the usual fate for a stony meteoroid of that size. Our explanation appeared in the scientific journal *Nature* in January 1993, two months before SL9 was discovered.

A second useful analog is preserved by the craters of Venus. Venus is wrapped in an atmosphere 100 times thicker than Earth's. Only large bodies can penetrate to crater the surface. This happens about once in a million years. Consequently Venus has only large craters. Small craters do not form because small impactors cannot get through the atmosphere. Venus has recently been mapped by NASA's *Magellan* spacecraft. The number of craters of different sizes on Venus offers a direct quantitative test of

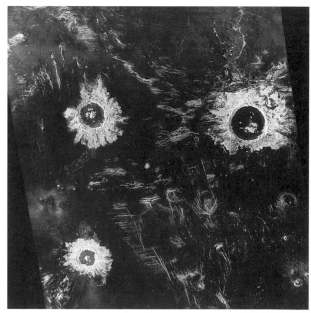

Venus's thick atmosphere (left) means that only relatively large impactors can penetrate to the ground and form craters. The Magellan radar image

(right) shows three large impact craters with diameters between 37 km and 50 km in the Lavinia Planitia region of Venus. (JPL/NASA)

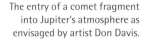

The entry of a comet fragment into Jupiter's atmosphere as envisaged by artist Don Davis.

how large a potential impactor must be to succeed in producing a crater or, equivalently, how effectively an atmosphere protects a planet's surface. In 1992 I developed a simple description of how kilometer-size impacting bodies interact with an atmosphere that allowed me to simulate the craters of Venus. With the help of my friends, this description evolved into the explanation of the Tunguska event and, at Jupiter, went from postdictive to predictive mode.

First predictions

We issued our first predictions at the June 1993 meeting of the American Astronomical Society. In the essentials, we would issue the same predictions today, given another doomed comet of unknown size and composition. We said (more or less verbatim):

> The comet strikes Jupiter's atmosphere at 60 km/s. It is promptly fragmented and fluidized and then flattened to a pancake by aerodynamic forces. The pancake intercepts much more gas than the original sphere. The gas halts the comet's progress. A 10-km-diameter comet with the density of water would reach the 100 bar pressure level in Jupiter's atmosphere, where the pressure is comparable to

that at the venusian surface. A 1-km comet would reach the 10-bar level. The comet stops suddenly, within a second or so. The net effect is that most of the comet's enormous kinetic energy is liberated in that second. A great deal of jovian 'air', as well as the stuff of the former comet, gets very hot and very compressed. Almost immediately thereafter the hot gases expand – violently. For all practical purposes the result is an explosion. Because the explosion is very large and occurs relatively high in the atmosphere, it does something unfamiliar, but long predicted in the violent world of blast wave theory: it punches a hole through the atmosphere and breaks out into space. The result would be an ejecta plume a thousand kilometers high and thousands of kilometers wide.

Two ways to make predictions

Over the course of the next year, predictions about the impact phenomena became much more refined and a lot more specific. Serious attempts were made to calculate what might be observable from Earth or from the *Galileo* spacecraft, which was more favorably located. There were two basic approaches. Analytic models use equations worked through with pencil, paper and home computer

Left: A numerical simulation of the entry of a 1 kilometer diameter sphere, with the density of water, into Jupiter's atmosphere. Color corresponds to density: the atmosphere is compressed (red) in front of the incoming projectile but forms a very low density wake (dark blue) behind it. The projectile has just passed below the 1 bar level in this simulation and has not yet broken up: it is the uniform pinkish-red patch in the middle of the image. The perfect symmetry of the wake betrays the fact that this is a two-dimensional simulation, which assumes that conditions are identical at all angles around the projectile.

Right: The same simulation at a later time when the projectile is disintegrating and releasing its energy to form the beginning of the fireball. This simulation has penetrated very deep, 48 kilometers below the 1 bar level: it is unlikely that the SL9 fragments really penetrated this far, perhaps because they were smaller, less dense, or more rigid than the impactor in this simulation. (M.-M. Mac Low and K. Zahnle)

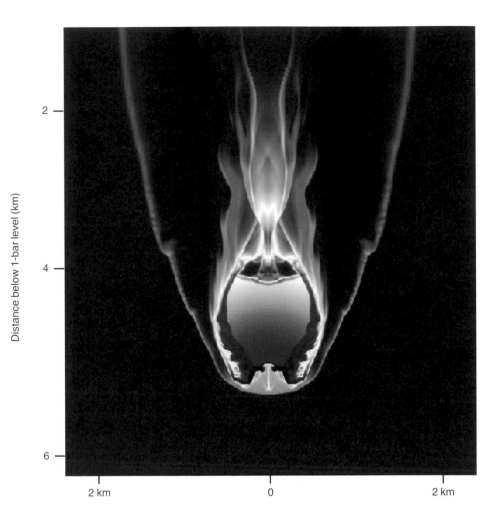

and are concerned with understanding the basic physics of what is going on. They lend themselves to sweeping generalizations. These appeared first and didn't change very much over succeeding months. Our Tunguska model is analytic. Numerical simulations require the power of big computers and are concerned with completeness and accuracy. Their purpose is to mimic reality. Numerical experiments are anecdotal.

If done correctly, the two approaches complement each other rather than compete, with progress in one leading to progress in the other. But if weighed as a competition, the analytic approach described the entry – the airburst altitude – more usefully, while the numerical simulations did better with the exit – the ejecta plume.

Analytic approximations are mostly about neglecting things. The ultimate goal is to arrive at a simple expression that retains only what really matters, while everything else is left out. The process begins with a simple picture of how things might happen. One constructs equations that

describe this simple picture. The equations let you follow the logical consequences of your thoughts. They must always be simple enough that you can go on to the next step. At most steps you find you have to leave something out in order to go on. If you leave out something that really matters, the equation you end up with is probably irrelevant. If you do get it right, you probably won't know it, and what you have found will be valid only within certain limits, which you only suspect. An analytic approximation, standing alone, is never truly convincing.

Numerical simulations begin with the fundamental equations of fluid dynamics. The physical quantities represented in these equations (temperature, pressure, velocity, and so on) are evaluated on a grid of thousands of discrete points that approximate the continuum of the real world, and are then evolved over time by various techniques, according to the particular code. Higher resolution models have a finer grid and more points to calculate. In principle this is about all there is to

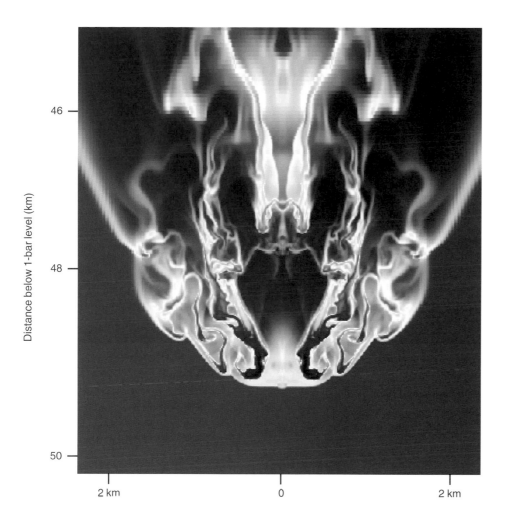

numerical models. The promise of simplicity and completeness is what makes computer models so compelling – and the videos they generate are so entertaining. In practice, numerical models are not so pure, nor so simple. A numerical model, standing alone, is just a pretty picture.

I am not a numericist by training or inclination. I had thought, as I presume you are now thinking, that provided the computers are coded correctly, the various different computer models would all give approximately the same answer. This faith proved naive. I learned better from my numericist colleague, Mordecai-Mark Mac Low of the University of Chicago. It turns out that contemporary computers are not nearly as powerful as one might gather from the popular press. No computer is large enough to perform a good simulation of a comet impact on Jupiter directly in a practical amount of time. Compromises must be made. The compromises determine the result to a surprising extent.

Compromises and approximations

One key compromise was that the entry calculation (following the comet into the atmosphere until it stops) and the exit calculation (following the explosion that occurs after all the energy of the impact has been placed in the atmosphere) had to be performed separately. That was because the scales of the two kinds of simulations are very different.

Entry simulations follow the deceleration and disruption of a kilometer-size body as it plummets into the jovian atmosphere. The experiments were typically performed in cyberspace cylinders several kilometers wide and tens of kilometers long. In practice the best way to do the experiment was to try to hold the comet steady and let Jupiter flow by. The more natural alternative, dropping the comet through hundreds of kilometers of jovian air, is computationally prohibitive. Exit simulations follow the dynamics of the explosion. The largest plumes on Jupiter rose over 3000 km high and extended over 10 000 km downrange. Very large, very coarse grids are unavoidable. In three dimensions the resolution is even worse.

There are other compromises. (The list here is not intended to be exhaustive.) The most important is resolution. High resolution is required to follow details in the destruction of the comet, but high resolution sorely taxes the machines. One can't follow as much of Jupiter as one would like. As a result the calculation is either coarse, or truncated so that important stuff slides off the screen into numerical oblivion. The better entry calculations are modeled in cylindrical symmetry (a two-dimensional calculation) rather than in full 3-D. The attempts at 3-D simulations of the entry are badly lacking in resolution.

The exit calculations, to be truly compelling, must be done in 3-D, but the computer requirements are such that these calculations are both coarse and few. The 2-D exit calculations assume that the comet falls vertically. In reality the SL9 fragments entered at a 45-degree angle. Consequently the geometry of the 2-D model is unrealistic. The extant 3-D plume models did a very good job of predicting the shape of the explosion and the shape of the ejecta pattern. On the other hand, the bilateral symmetry inherent in a 2-D model tends to make the simulation more evocative to the human eye, and the numbers are more tractable to analysis, so I am rather partial to 2-D.

An unfortunate consequence of the computational division into two kinds of numerical experiments was that far too much attention was focused on the entry simulations. This overemphasis on the less interesting problem was in part due to competition between the numericists, and in larger part due to journal editors, who were fixated on setting events in chronological order. Whereabouts a well-characterized impactor will explode is indeed an important problem. Our interest in this problem is not limited to impacts on Jupiter. It also involves a healthy respect for our own self-interest. Stray bodies can and will hit Earth, and we should know what to expect, and what evasive action, if any, we should take. But for SL9 the emphasis was misplaced. We had then, and have now, very little direct evidence of just how large the SL9 fragments were. In the end, all the numerical groups spent inordinate amounts of time treating the entry problem, writing and rewriting the same basic papers to satisfy journal editors who

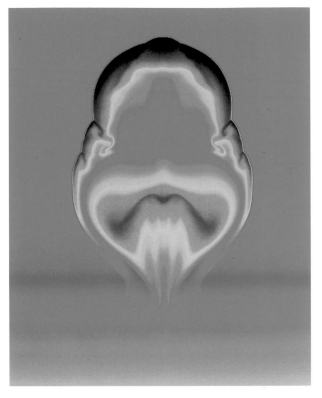

A supercomputer simulation of the explosion of a comet fragment in a region of jovian atmosphere 1000 km across and 1200 km high. The darkest blue horizontal stripe is roughly where Jupiter's visible clouds are. The colors represent a scale of temperature. Most of the vaporized comet material is in the dark orange 'moustache'. The much hotter shock wave above is pure jovian 'air' but the visible fireball is lower than the shock wave because the comet material is more opaque. (M.-M. Mac Low & K. Zahnle)

demanded certainty where none was to be hoped for, and who, ultimately, didn't publish the papers anyway.

Pre-impact hopes and fears

Before the impacts took place there was a widespread fear among the *illuminati* that the events would not be detectable from Earth. Geometry was against us. The fragments would strike the far side of Jupiter as viewed from Earth, so observing an impact directly would be impossible. Meanwhile the *Galileo* spacecraft, which was well placed to watch the impacts, was (and is) so crippled by a feeble transmitter that weeks would elapse before we'd find what, if anything, it saw.

This fear was amplified by extensive coverage in the mass media. Optimists spoke freely about the enormity of the coming events, while pessimists mumbled darkly about Kohoutek, 1973's 'comet of the century' that proved mostly a dud. My reading of the community is that, before the event, the general viewpoint was a kind of open-minded pessimism: no one really expected to see much, but because the payoff would be so high, most were betting on optimism.

Thomas J. Ahrens and his colleagues at Caltech were the first to realize that, although the impacts would indeed occur on the far side, the ejecta plumes would rise above the limb and so become visible from Earth. The competing groups quickly made their own calculations of the same effect. There was a great deal of debate over just how visible the plumes would be when they popped over the horizon – whether they would be so cool and thin that they would glow but faintly, or scarcely reflect any sunlight at all. Still it did seem distinctly possible that the rising plume would be visible from Earth about a minute or so after a fragment struck the planet. This prediction in its essentials was eventually borne out. The rising plume was distinctly detected by Earth-based telescopes as an infrared flash during several of the impacts, though it was subtle and completely overwhelmed by later events. It was also beautifully imaged by the Hubble Space Telescope.

The best prediction, and still the best estimate for the size of the fragments, was based on models of the tidal disruption of the parent body (see Chapter 7). These models show that the largest fragments were probably about 500–700 m across and had densities effectively about half that of water. We had predicted that bodies of this size and density would explode at about the 1-bar level, i.e. where the atmospheric pressure is comparable to that at Earth's surface. Ultimately, the sulfur compounds generated by the impacts and detected from Earth suggested that the largest fragments reached Jupiter's possibly sulfurous clouds at 1–3 bars but did not get deep enough to exhume water. However, if the sulfur originated with the comet, as is possible, its presence is no indicator of the depth of penetration, which remains the subject of debate.

Several groups hoped that the impacts would tell us not just about impact phenomena, but about Jupiter itself. They reasoned that the impact explosion should produce at least two types of waves. Seismic waves would spread rapidly through Jupiter's interior and, with luck, would return to the surface after reflection from boundaries between different regions, such as the transition from molecular to metallic hydrogen. If the returning waves heated the upper atmosphere sufficiently to show up in

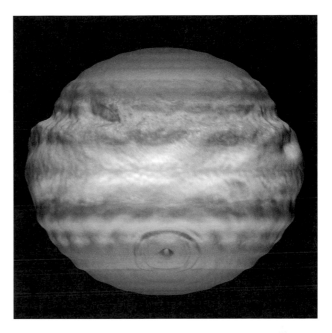

It was hoped that the impacts would generate visible gravity waves (that is, waves where gravity pulls the material back to its original position after the wave has passed, such as ordinary ocean waves). In this computer simulation of Jupiter's atmosphere done by scientists at Massachusetts Institute of Technology (MIT) and published in *Nature*, atmospheric pressure is shown as exaggerated surface relief, so high pressure regions appear physically high. The computer has simulated an impact in the southern hemisphere, and gravity waves spread out from it like ripples in a pond. The dark expanding rings seen around many impact sites by the Hubble Space Telescope may have been a variant on the expanding rings the MIT group predicted. (J. Harrington, R. P. LeBeau, K. A. Backes and T. E. Dowling, MIT)

the infrared, we might learn the depth of this or other transition regions. Theoretical models would be put to an unexpected observational test. Slower waves should spread through the atmosphere, revealing by their speed fundamental data on jovian atmospheric conditions.

The hot surprise

The big surprise was that many of the impact events were easily visible and long-lived (typically 20 minutes or so), provided one was equipped with the heat sensors to detect them. The fireworks were exclusively an infrared affair. In visible light there was little to see but the ashes.

As the impact sites rotated into view they were revealed as strong sources of infrared light. The net effect was spectacular and unexpected. Detectors were saturating worldwide. The theorists had said nothing about this, or at least nothing intelligible. They had some explaining to do. The theorists had said many things, of course, many of which were contradictions. We were aware that the ejecta would carry a great deal of energy, and would be broadly distributed. We said so, often. But for reasons which I can no longer reconstruct, or even imagine, it didn't occur to us that there would be dust present in the ejecta. So nobody calculated how much infrared light would be emitted by warm dust when the ejecta fell back. I remember that I was fixated on visible light, probably as a result of watching too much television in my youth. This inability to reconstruct the pre-impact predictive climate is the chief peril of trying to write about predictions six months after the events have taken place.

In hindsight, the story of the infrared light curve is now pretty clear. The best events observed from Earth were characterized by three flashes, the first two faint and brief and the third spectacular and prolonged. The first flash was directly associated with the entry of the comet into the atmosphere. Why it could be seen from Earth remains a puzzle, since the impacts did indeed

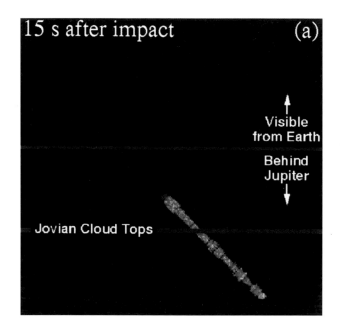

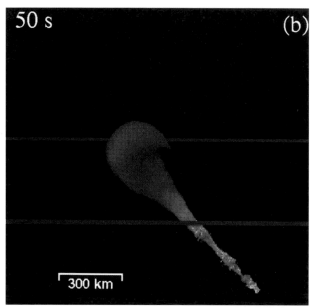

A numerical simulation of the fireball's exit along the entry track, published on July 5, 1994. The authors' hope that the fireballs would rise high enough to be visible above Jupiter's limb was spectacularly realised 11 days later. (Mark Boslough, D. Crawford, A. Robinson and T. Trucano, Sandia National Laboratories)

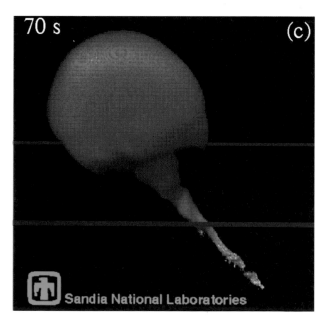

occur on the far side of the planet, as expected. One possibility is that the light from the explosion somehow reached Earth by scattering off of high altitude dust, presumably associated with the impacts. The dust could be from previous impacts, from the coma and tail of the impacting comet itself, or it could be dust formed or deposited in the high altitude meteor trail. A second possibility is that we saw directly from Earth the light from the high altitude portion of the meteor trail. The latter I think more likely, especially as parts of the flash seen from Earth appeared to precede the moment of impact, but the matter is unsettled.

The second flash corresponded to the fireball rising above Jupiter's limb. This flash was predicted by several groups. We may deserve some credit for getting this one right; in any event, we will take credit.

The third flash, much the brightest and longest lasting, corresponded to the ejecta plume falling back down onto the atmosphere. A loose analogy is what happens when one drops a large rock into a lake. The rock kicks up a fountain of water (the ejecta plume) which then splashes back down. The plumes were imaged directly by the Hubble Space Telescope. They typically rose 3000 km above the cloudtops. When they descended, they fell those same 3000 km back down again before hitting the atmosphere. It took 10 to 20 minutes for the plume to rise and fall again, by which time the impact site had rotated into view from Earth. A huge amount of energy was released over a very large region when the plume fell. Because the area was large, the emission appeared bright from Earth.

Predictions were better met by *Galileo*'s observations. Because *Galileo*'s view was unobstructed it saw but two peaks, one corresponding to the rising, cooling fireball and the other corresponding to the re-entry of the plume into the atmosphere. Our group had predicted that the entry of the meteor and the peak brilliance of the fireball would be separated by tens of seconds. This proved mostly wrong. The alternative view, put by a group working at the Sandia National Laboratories in Albuquerque, New Mexico, – that the entry should be regarded as a line charge, with the wake exploding while the comet continued its descent – turned out to be mostly correct.

One of the nicest verifications of a prediction was the beautiful correspondence between HST images of the plume and a 3-D calculation by the Sandia group (see Chapter 9). Several numerical experiments, by both the Sandia group and the Caltech group, showed that the explosion was channeled up the wake. Material was preferentially ejected along the path by which the comet entered. This occurred because the wake is filled with a hot, low density gas. The speed of sound is high in this gas, which means that the shock travels fastest in the wake. Because of the low density, there is less material to move out of the way. Both factors make the wake a highway to space.

Not all predictions were ambiguous, lost in the noise, or reinterpreted to fit the outcome. One published prediction (Zahnle *et al*, EOS, 1993) I am happy to recall. 'The removal of this comet leaves the solar system a slightly safer place for us all.' Indeed it did. But it doesn't feel that way, does it? The solar system no longer seems quite so far away as it did before July, 1994. Here we are, close to the edge, protected from the true enormity of the universe by a thin blue line. A day will surely come when the sheltering sky is torn apart with a power that beggars the imagination. It has happened before. Ask any dinosaur, if you can find one. This is a dangerous place.

While no-one knew what the effects of the impacts on Jupiter were going to be, some scenarios, such as this one, could confidently be ruled out! (John Westfall, Association of Lunar and Planetary Observers)

6

The International Observing Campaign

Michael F. A'Hearn

With just about a year to do it, astronomers set about laying plans to monitor Jupiter during Comet Crash Week. The power of electronic communications meant that information could be disseminated quickly and freely. Within moments of the first strike, the surprise news of the fireworks sped round the world and observers adapted their plans accordingly.

In May 1993 astronomers realized comet P/Shoemaker–Levy 9 would strike Jupiter. It was immediately obvious that this once-in-a-millennium event demanded an extensive and coordinated observing campaign. As it turned out, the unprecedented nature of the impact, and its timing, led to major changes in the sociology of astronomical research. A totally new kind of campaign was launched.

It was clear that the two key ingredients of a successful program would be telescope time and funding to get equipment to the telescopes. Attempts to organize a program began in the USA, largely because there are as many active planetary astronomers in the USA as in the rest of the world put together. Jürgen Rahe at NASA and Morris Aizenman at the National Science Foundation (NSF) led the early efforts. Very soon Richard West at the European Southern Observatory's headquarters in

Germany was taking a similar lead. These were the first formal, large-scale initiatives but many other observatories and agencies rapidly joined in. Work on coordination, securing funds, and getting telescope time involved many people whose names are not mentioned here. There was no clear division between the various activities as everyone involved did all they could to put in place the most ambitious astronomical observing campaign ever undertaken.

Raising the money

In the (northern) summer of 1993, I and a number of other representatives of the research community and the funding agencies in the US estimated the total cost of a realistic program for US participants. They would need money particularly to meet the costs of travel to the southern hemisphere and for additional equipment. The two funding agencies in the USA, NSF and NASA, recognized at once the importance of the proposals. Acting faster than on any previous occasion, they changed their priorities to create a special program of grants. It was a testimonial both to the determination of individuals in the two agencies and to the widespread scientific interest in the impending impacts.

Our back-of-the-envelope estimate of the cost of a proper program turned out to be within 25% of the total amount of money actually requested from NASA and NSF. Of course, despite tremendous efforts, the agencies could not meet all requests. The total amount of money ultimately made available in the combined NSF/NASA program was about $1.25 million. Significantly, it was the first time that NSF and NASA coordinated their grant allocations for planetary astronomy. The normal pattern was to consult informally about overlapping proposals but, in the end, they would make their decisions quite independently. In this case, however, the two agencies each convened panels to review the applications for funds with many members in common so they could distribute awards fairly and sensibly.

Similar exercises were taking place all round the world to redirect resources for a unique astronomical event that would not wait while the bureaucrats took their time.

Coordinating observations

To tackle the question of telescope time allocation, we organized a tele-conference in mid-1993 between representatives of NASA, NSF and all the observing facilities funded by the two agencies. All agreed that observations of the impacts would be scientifically important. All were alerted at this early stage to the fact that there might be a rush of applications, all wanting use of a given facility at precisely the same time. Furthermore, all agreed that, because of the unique character of the events, they would make the data publicly available very rapidly in order to stimulate comparative studies.

Many observatories modified their normal processes for reviewing proposals. For example, the Space Telescope Science Institute (STScI) allocated around 100 orbits of the Hubble Space Telescope, equivalent to one fortieth of the total time available in 1994, to the comet impact programs. A special committee was convened to review the applications and it arranged 'shotgun marriages' between certain competing proposals. NASA's Infrared Telescope Facility (IRTF) in Hawaii set aside a month of observing time and also convened a special

committee. A team was selected to make 'family decisions' on how to carry out the program.

Planning

Jay Melosh at the Lunar and Planetary Laboratory of the University of Arizona, Tucson, organized the first workshop about the comet and Jupiter in August 1993. Originally intended to be just a small workshop for local people, this meeting drew more than 100 participants from across the USA, including both observers and theorists. It was the first quantitative indication of the widespread interest in the phenomenon. This workshop focused on the scientific issues as they were then understood. Already the variety of predictions was enormous. Some were saying that all the energy would be released above the clouds of Jupiter resulting in spectacularly bright flashes while others suggested that all the energy would be deposited well below the clouds and very little would be observable from Earth.

The Tucson workshop was primarily to discuss the scientific questions. The next workshop, a one-day session in November 1993 organized at ESO headquarters by Richard West, was the first of several aimed at putting together a specific observing strategy for a particular facility and at obtaining the telescope time to carry it out. Most of the interested groups of European planetary astronomers were represented at this workshop. Participants assembled a comprehensive plan and submitted it for approval. The allocation committee combined the few proposals already in its hands with the large package from the workshop and dedicated a very substantial block of observing time to the study of the impacts, involving every telescope at its disposal.

Lucy McFadden and I organized a two-day workshop at the University of Maryland in January 1994. Roughly 165 people who were planning to participate in the campaign attended. We wanted to use the predictions made by theorists to optimize and coordinate the observational programs. The range of predictions had not narrowed since the previous August but the debates surrounding competing theories were based on more detailed calculations. By this time most major facilities had selected their observers and/or observing teams so individuals were able to take advantage of this meeting to decide how best to utilize their observing time in view of what would be done elsewhere.

Another concern was to make observations at precisely the wavelengths where we expected to see interesting phenomena. This required special filters, tuned, for instance, to the wavelength bands in which the methane in Jupiter's atmosphere absorbs. John Spencer and Glenn Orton solicited, gathered and collated filter requests. They communicated with potential observers through the electronic mail network of the International Jupiter Watch, an informal group of professional Jupiter observers that had existed since 1987, and by means of the electronic bulletin board set up specially at the University of Maryland in anticipation of the comet impact.

Many other meetings took place, including a special scientific symposium in Japan in November 1993, and special sessions at the regular meetings of the American Astronomical Society's Division for Planetary Sciences in October 1993, the American Geophysical Union's December 1993 and May 1994 meetings, the American Astronomical Society in January 1994, and the Lunar and Planetary Science Conference in March 1994. There were, of course, innumerable meetings of teams from various large facilities for planning purposes.

The diversity with which different observing teams tackled the challenge was remarkable. Some people planned on using a single telescope with a single instrument, dedicated to their use. Others organized large teams with portable equipment to disperse to a series of sites at different longitudes. Some individuals associated themselves personally with programs at up to a dozen major observatories. Considering the 'shotgun marriages' at some observing facilities, it was truly surprising how few divorces ensued.

Communications

It was obvious at an early stage that rapid communication among observers would be an essential ingredient of a successful observing campaign. One of the outcomes of the January workshop in Maryland was a consensus that a combination of two techniques was desirable. One element would be an electronic bulletin board for messages that were long or considered less urgent. Users would be able to scan the bulletin board at will from their own computers. The other communication tool would be an e-mail 'exploder', which would automatically distribute short messages rapidly via participants' computers, without manual intervention. The Small Bodies Node of NASA's Planetary Data System (PDS-SBN), which is located in the University of Maryland's Department of Astronomy, was

commandeered to perform these functions. Its normal job is to archive data from NASA planetary exploration missions and to make such data, and other information about the small bodies of the solar system, available to researchers. Anne Raugh set up the software for the communications and monitored the traffic. We limited the list of recipients for exploder messages, primarily to active observers, although others were included if we thought it would be generally beneficial to do so. We advertised the bulletin board only within the research community and initially just researchers and a few scientific journalists used it.

In May 1994, access to the electronic bulletin board was extended by means of the World Wide Web (WWW). Internet users can easily track down astronomical data worldwide using a simple graphical interface to reach many different participants in the WWW system. We continued to limit access to the exploder. Even so, we had approximately 250 recipients on our distribution list, representing every continent and 25 countries.

Before the week of the impacts, most communication activity consisted of people taking information from the bulletin board, as we had expected. Data available included predictions of the impact times and places, predictions of phenomena, ephemerides and orbital elements, historical information, lists of papers and of planned meetings, etc. The bulletin board served as an ideal location to post helpful graphic images. These graphics included numerous diagrams illustrating the geometry of the impacts, provided and regularly updated by Paul Chodas and Don Yeomans. There were other selected graphics, such as diagrams showing the Jupiter-facing hemisphere of Earth at the time of each impact (also updated several times) from Larry Wasserman. These graphics were invaluable as planning aids and the willingness of astronomers to share their work was impressive. We even had specially written software posted on the bulletin board.

The statistics on the numbers of people retrieving information through the computer network, and when they were doing it, reveal interesting patterns of activity, as well as demonstrating the popularity of the service. Over the first two weeks of July, as people planned their observing sessions, there was a steady upward trend in the numbers of users. The average was 58 each day. Most were taking Sundays off, but not Saturdays. There were about the same number of connections to the World Wide Web but the people using it were taking off both

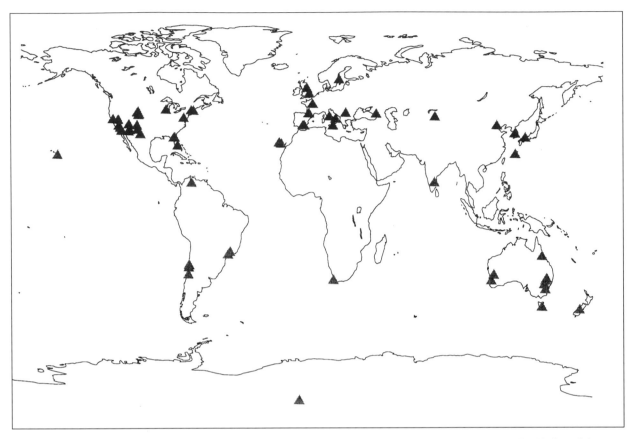

Map showing all ground-based sites, amateur and professional, that reported observations of the impacts and related phenomena to the University of Maryland e-mail 'Exploder'. Many reports were also received from spacecraft, ranging from the Hubble Space Telescope 590 km above the Earth, to *Voyager 2*, 8.3 billion km away, beyond the orbit of Pluto.

Saturdays and Sundays, suggesting perhaps that they were accessing the system from their workplace but were not active participants in the observing campaign. During that same period, we averaged two messages per day sent through the exploder. At this point we instituted 24-hour live supervision of the bulletin board in order to be sure that everything we received was very promptly posted. Lucy McFadden was responsible for the astronomical aspects while Anne Raugh was in charge of the computing side. Numerous Maryland students were pressed into service to cover the night shifts.

Impact week started on Saturday 16 July. In an article published in the scientific journal *Nature* a couple of days earlier, and noted in an exploder message, Paul Weissman had pointed out that his 'rubble- pile' model for the comet predicted a bright flash but relatively little by way of post-impact consequences.

We received the first report of an observable effect on Jupiter at our bulletin board at 23:29 Universal Time (UT)

on Saturday 16 July from Tom Herbst and his collaborators at Calar Alto in Spain, who observed an infrared flash. With hindsight we now think it was due to the collapse of the A impact plume after it had rotated into direct view. Anne Raugh promptly posted the message to the exploder.

The number of messages to the exploder itself exploded, soaring to 13 on the 16th, 51 on the 17th, and 62 on the 18th. By late on the 16th we received from Ulli Kaüfl the first infrared light curve, which was for Fragment A. Since A was one of the dimmer fragments, and it produced very bright flashes in the infrared, it was already clear that this was going to be a week of spectacular fireworks, although the messages to the exploder and to the bulletin board were still reasonably restrained.

By early on the 17th we had posted images to the bulletin board and were receiving reports from spectroscopists about what chemicals had or had not

Part of the 'Home Page' of the Jet Propulsion Laboratory's Internet 'World Wide Web' site for SL9 information, as it appeared on 21 January 1995. The World Wide Web software turns the Internet into a global multimedia encyclopedia and it was a very popular means of obtaining information and images about the comet crash during impact week. Clicking with a computer mouse on the blue-coloured text provides access to additional information, images, or animations on that subject, often obtained from another location that can be anywhere in the world. (JPL Homepage managed by Ron Baalke, JPL)

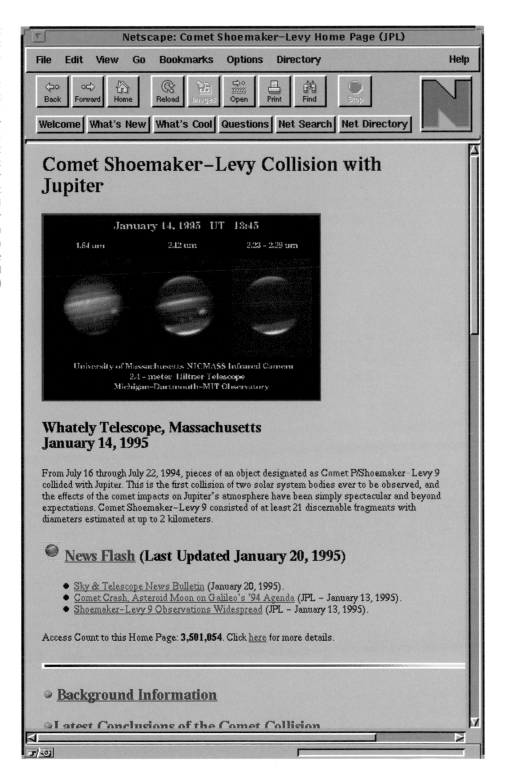

been detected. By the evening of the 17th people were publicly puzzling over the data, with John Spencer, who was in Chile, wanting to know how the spots got so big so fast.

After the impact of fragment G, the brightest fragment to hit Jupiter to that point, there was no more restraint. Hien Nguyen exclaimed from the South Pole, 'My God, it was extremely bright!', a statement criticized as a possible understatement. In Australia, Vikki Meadows related how they had been compelled to stop down the main mirror of the Anglo-Australian Telescope from 3.9 m to only 1.9 m because the brightness had saturated their infrared detectors. The rate of messages through the exploder got so high that some individuals who were not totally involved requested to be deleted because they could not keep up with the mail.

By this time we were also seeing interpretations of the results flying back and forth, including estimates of the depth to which the bolide must have penetrated and of the altitude of the clouds. We also began seeing the requests for critical timing data to help the *Galileo* teams decide which data to transmit back to Earth. Interestingly, at this point it was still quite unclear whether any of the observed phenomena were correlated with, and could thus give the time of, the actual impacts. This question was not fully resolved till many months after the event.

The 18th was the day on which the exploder messages peaked at 62 and also the day on which usage of our bulletin board peaked. Also on the 18th we had an average of more than 1000 accesses per hour via WWW. This was halved by the 20th, never again to rise above that level. The number of different places accessing the bulletin board by WWW peaked at 1100. The middle of impact week was also the time at which several other sites with WWW information completely choked up with traffic so that no one could access the data through WWW. Fortunately, we at Maryland were spared this crisis. Active participants in the campaign subsequently reported to us that ours was the only site that was sufficiently uncongested that they could retrieve images. From this point on, it was all downhill – not because

exciting data were no longer coming in but because most people have trouble sustaining such a degree of excitement for five days and enough of the pattern of phenomena was known to make communication somewhat less urgent.

The real benefit of the rapid communications was twofold. First, many people changed observing strategies during the week in response to reports over the exploder. Having the most recent predictions for the times of the impacts was important, not only for deciding when to concentrate one's attention on the stream of data but also for understanding the significance of what other people had seen. This was certainly true in our own case observing from Western Australia. The lack of any detections of flashes reflected from the jovian satellites led us to change the gain and offset on our high-speed photometer while the reports of visible spots led us to change the choice of filters for our CCD system. We presume that many other observers made similar changes.

The second benefit was an indirect one. All the data were widely available, which resulted in global coverage by the media. The rapid growth in the use of WWW over the preceding year or so, coupled with the recognition that rapid communication was essential, made this possible. Many institutions set up WWW sites specifically to disseminate images as widely as possible for those outside the observing campaign. Many of these sites were themselves peripheral to the campaign but they were able to take advantage of the fact that astronomers rapidly posted their data in places that are accessible electronically. While some may have intended mainly to help observers, others certainly wished to broadcast the data as far as possible.

7

The String of Pearls

Harold A. Weaver & David Jewitt

At discovery, the fragments of Comet SL9 were already strung across the sky, rather like a pearl necklace. As the inevitable crash drew nearer, the train of nuclei lengthened and changed in appearance. Some pieces vanished while others broke up further. Were they really the remains of a comet, or had they once been part of an asteroid?

Comet P/Shoemaker–Levy 9 (SL9) ranks as one of the most unusual astronomical objects ever observed. Astronomers have recorded many split comets in the past, but never one so dramatic as SL9. In the typical splitting scenario, there is one dominant fragment and a few others are observable for at least short periods. SL9 broke into about 20 observable pieces, most of which remained visible over the entire lifetime of the comet. In addition, a majority of the fragments appeared to lie along a single straight line projected on the sky, prompting observers to

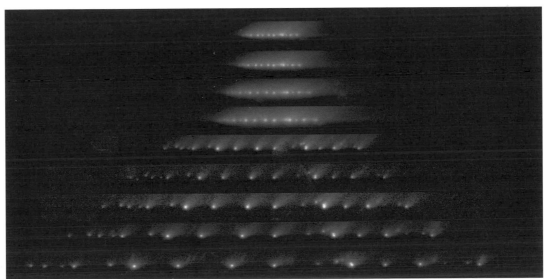

27 March 93
15 April 93
12 June 93
17 July 93
14 January 94
14 March 94
13 April 94
8 May 94
7 June 94

This series of CCD images of SL9 was taken with the University of Hawaii's 2.2-m telescope at Mauna Kea Observatory between discovery and one month before impact. They are all to the same scale and show how the train of nuclei lengthened over that period. The fading of the dust sheet between the fragments from 1993 to 1994 is clearly seen, but it appeared that all fragments maintained their individual dust clouds throughout this period, with the exception of J and M which either lost their dust or disappeared entirely. Note also the changing apparent positions of the fragments, keeping in mind that we are seeing a projection in two dimensions of their actual three-dimensional arrangement in space. (D. Jewitt & J. Luu)

dub SL9 the 'String of Pearls' comet. This unusual configuration persisted through the entire apparition of SL9, although the length of the 'train' (another term commonly used to describe the aligned fragments) increased dramatically. At discovery in late March 1993, the apparent length of the train in the sky was about 50 arcseconds. By mid-May 1994 it had increased to about 360 arcseconds. Over the short interval from mid-May 1994 until the impacts in mid-July, the train length grew even faster, reaching around 900 arcseconds as the first fragment plunged into Jupiter.

Appearance

The basic morphology of SL9 changed slowly but significantly during the 16 months of pre-impact observations. At the time of discovery, in March 1993, the component fragments were embedded in a sheet of bright material that extended beyond the ends of the train. Most of the light from the comet seen at this time came from the sheet. The embedded nuclei made only a minor contribution. Charge coupled device (CCD) spectra showed that the light from the sheet was reddened sunlight, presumably reflected from innumerable particles of reddish or brownish dust and larger pieces of solid material. Most likely, the dust sheet consisted of debris ejected at the time the comet broke up and

distributed in SL9's orbital plane about Jupiter. This material faded progressively with the passing months, and by the time of impact it was a mere vestige of its former self. Either material in the sheet was not replenished or it was replaced too slowly to compensate for the loss suffered in the spreading process. Solar radiation pressure eventually blew away the small dust particles produced during the tidal break-up of SL9. As happens with any ordinary comet, radiation pressure was also responsible for producing the dust tails observed in SL9 and for keeping most of the dust sheet on one side of the train. That was the side in the direction away from the Sun at the time of breakup.

As the dust sheet faded, the embedded components changed in subtle ways. In addition to the slow lengthening of the train, most fragments developed aligned dust tails, which persisted until shortly before their impacts on Jupiter. The relative brightness of the component parts changed on timescales of weeks and longer. Some brightened and others faded, in some cases until they became invisible. Several fragments that originally looked like single objects became clearly resolved into several pieces as they moved apart from one another over time and/or were observed with higher resolution telescopes. During most of the observing period, each fragment had its own inner coma (dust

By 16 February 1994 SL9 was fainter and much more spread out than in early 1993. In this image, taken with the 0.91-m Spacewatch Telescope at Kitt Peak, dust tails are clearly seen streaming back from each nucleus but the 'wings' that had extended beyond the train have faded and are no longer visible. (R. Jedicke, T. Metcalfe, J. Scotti)

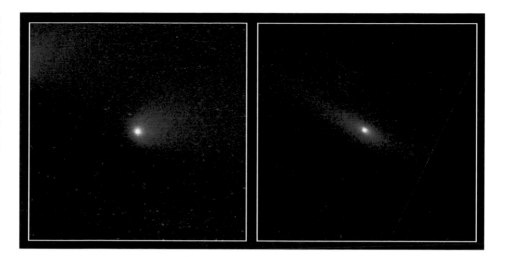

The H nucleus as viewed by the HST in January and July 1994. In January it had a spherical inner coma and a dust tail pointing away from the Sun, more or less like a normal comet. By 14 July, tidal forces due to Jupiter had stretched the coma so that it was elongated both towards and away from Jupiter. (H. A. Weaver & T. E. Smith, STScI/NASA)

SL9 as seen from Pine Mountain Observatory, Oregon, on 9 July 1994. The comet fragments are visible as five luminous condensations in a more diffuse train. The diffuse glow in the top left is scattered light from Jupiter. (G. Bothun/University of Oregon)

cloud), which appeared circular and was about 1 arcsecond across. In the final few weeks prior to impact, however, the comae became highly elongated along the comet–Jupiter line as Jupiter's tidal forces again exerted significant influence. The comae took on a tadpole-like appearance, with the tail pointing toward Jupiter. Despite the stretching of their comae, the fragments generally maintained condensed, stable cores, at least until the last views about 10 hours before impact.

Nomenclature

The numerous components of SL9 presented a problem of nomenclature. After originally using a numbering convention for the fragments that started at the south-west tip of the train (with fragment number 1) and moved toward the north-east tip (fragment number 21), most astronomers finally adopted a naming convention using letters. The first fragment to hit Jupiter (on 16 July 1994) was named the 'A' nucleus, while the last to hit (on 22 July 1994) was the 'W' nucleus. The letters I and O were not used because of the possible confusion with the numbers one and zero. Notice that the letter and number naming conventions go in opposite directions: for example, fragment A was also named 21, while fragment W was number 1. Some of the nuclei split further and the new fragments were given both primary and secondary

The last HST image to show the entire train of fragments, taken on 17 May 1994 with the telescope's wide-angle camera. The comet extended over 1.1 million km and the dust tails had faded significantly since the previous January. The nomenclature adopted for the fragments has been superimposed. (H. A. Weaver & T. E. Smith, STScI/NASA)

labels. For example, the Q (= 7) nucleus became the Q1 (= 7a) and Q2 (= 7b) pair.

The apparently linear arrangement of the SL9 fragments can easily be explained. When the comet first broke up, the fragments had essentially the same velocities with very little difference between them. Then they started to diverge, each travelling along separate, highly elliptical orbits. The orbits were different because of the different initial positions of the fragments relative to Jupiter. Nevertheless, all the pieces appeared to remain in a straight line as seen on the sky.

Off-train nuclei

While the brighter SL9 nuclei tended to lie on the train,

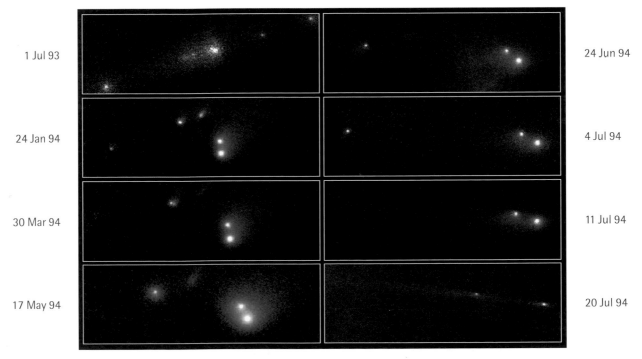

HST images of the region of the comet centered on nuclei P and Q, nicknamed 'the gang of four'. The sequence shows marked changes between July 1993 and July 1994. All frames are to the same scale. On 1 July 1993 the double nature of P and Q is barely detectable. By 24 January 1994, and with the help of HST's improved optics, the two components of each can be seen easily. By 30 March, fragment P1 had faded to a loose puff of dust. Meanwhile, P2 had split into two pieces, though only one of these survived in the end. The whole train continued to stretch out and the tails align towards Jupiter so that, by 20 July, 10 hours before impact, only Q1 and Q2 are visible on this scale. (H. A. Weaver & T. E. Smith, STScI/NASA)

ten of the fragments, particularly the fainter ones, were slightly displaced by amounts ranging from about 1 to 5 arcseconds. Although these displacements were small in relation to the length of the train, they were highly significant and need an explanation. At least two ideas have been suggested. One is that the off-train nuclei were produced by fragmentation events after the break-up of the original SL9. The other suggestion is that non-gravitational forces of some kind, such as the emission of jets of gas, pushed the smaller nuclei off the train. Examination of SL9 images strongly supports the conclusion that some of the SL9 nuclei fragmented well after the time that the main comet was itself split apart by its tidal interaction with Jupiter. For example, analyses of their orbits indicate that the Q1 and Q2 fragments probably split from a common parent in April 1993. The P2 nucleus became resolved as two separate objects by March 1994, and a small companion was found near the G nucleus in May 1994. These later fragmentations may have been due, at least partially, to internal forces arising from the rotation of the nuclei. If this were the case, the products of the splittings would have moved away from their parents with some small, but non-negligible, velocity that would then create the displacement from the train.

In order for the second explanation (i.e. the non-gravitational forces) to be viable, the nuclei of SL9 must have produced gas and dust continuously like a typical comet. However, it is not clear that there was any such continuous activity in SL9.

The fragmentation continues

Although new splitting events created extra nuclei, some of the fragments became 'missing bodies' as they disappeared from sight. Fragments J and M were observed in ground-based images shortly after SL9's discovery in March 1993 but neither could be detected at any time in 1994. P1 and T looked like little more than puffs of dust during early 1994 and both became undetectable in Hubble Space Telescope (HST) images taken in late June 1994. Perhaps the J, M, P1, and T nuclei fragmented into clouds of dust particles that ultimately diffused away under the influence of solar radiation pressure.

The appearance of new fragments and the loss of others clearly demonstrate the incredibly fragile nature of SL9. Even the tidal forces that split the original body were tiny, the stress being less than 1/100th of an atmosphere. A person could easily have pulled SL9 apart

with his or her bare hands. Since the tidal stress exerted by Jupiter declines rapidly with increasing separation, the comet's tidal interaction with the Giant Planet was probably not the cause of the later fragmentations. As when any cometary nuclei split, we do not really know why the SL9 nuclei continued fragmenting. Rotation of the nuclei would produce some structural stress. Perhaps this, coupled with an outburst of gas and dust, could fracture some nuclei. In any event, the forces responsible for the later splittings were probably smaller than the stress that split the original SL9 body. This demonstrates that at least some of the fragments were even more loosely bound than the original comet.

While the SL9 fragments were clearly very fragile objects, it seems unlikely that they were simply 'traveling sandbanks', consisting of swarms of thousands of similarly sized objects. The break-up of a single, coherent body (i.e. a nucleus) located near the center of the coma explains the splitting of one fragment into two distinct fragments most naturally. In contrast, a swarm would probably evolve into an increasingly diffuse cloud. Perhaps P1 and T ultimately disintegrated in this fashion.

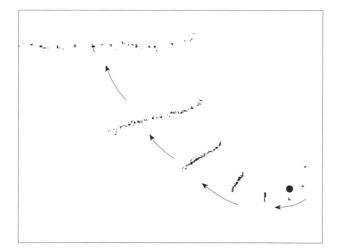

A computer simulation by Eric Asphaug and Willy Benz of the break-up of SL9 during its close approach to Jupiter in July 1992 and its subsequent elongation. It shows schematically how the comet would spread over time if it were made of a large number of pieces held together only by mutual gravitational attraction. As the comet passes close to Jupiter, tidal effects overcome the mutual attraction of the particles and they are stretched out into an elongated swarm. However, as the swarm recedes from Jupiter, neighbouring particles fall together again and the swarm naturally divides into about 20 clumps due to its own gravity. This model therefore predicts that each fragment of the comet was composed of many smaller pieces held together by gravity. The idea nicely explains the comet's separation into 20 well defined fragments but cannot easily account for the way the pieces continued to break up while a long way from Jupiter.

The highly tenuous structure of SL9 supports the idea that it was a 'rubble pile', consisting of a loosely bound agglomeration of many sub-units that came together when the comet formed.

How big were the nuclei?

It proved impossible to tell the sizes of the nuclei by telescopic observation. Not knowing reliable sizes for the fragments is probably the single biggest cause of uncertainty in the interpretation of the SL9 impacts. Uncertainty in the radius of a nucleus amounting to a factor of five, for example, implies uncertainty by a factor of 125 in the mass and energy of the impactor. Such doubt severely undercuts attempts to interpret observations of the impacts on Jupiter. As is typical of active comets at all distances from the Sun, light scattered from the nucleus itself is very difficult to separate from that of the enveloping coma. Owing to their superior resolution, HST images of the comet were thought to offer the best chance of determining the sizes of the nuclei. Although HST's resolution at the distance of SL9 was no better than 250 km, far larger than the sizes of the nuclei, it was hoped there would be a 'spike' in the observed brightness near the center of each coma that would represent a clear signature of the nucleus. Unfortunately, nature was not very cooperative. In the brightest pixel, the contributions from the nucleus and coma were apparently of comparable magnitude. Subsequent attempts to 'deconvolve' the HST images indicate that the larger nuclei were about 3 or 4 km in diameter (if the reflectivity, or albedo, of the nuclei was 4%), but the more conservative interpretation of the HST images is that these numbers are only upper limits to the true nuclear sizes and that significantly smaller sizes cannot be ruled out.

It is not even obvious from the images that the brightest components of the nuclear train were the largest underlying nuclei. Detailed observations of other short period comets show that the nuclei are mantled with material that inhibits the escape of gas except from localized active areas or 'vents'. On Comet Halley, for example, the vents occupy about 10% of the nuclear surface, while on Comet Encke the activity is confined to only 0.2% of the total surface. As might be expected, the coma brightness is correlated more closely with the active area than it is with the size of the nucleus. Observations of the impacts showed that the size of the impact phenomena were not perfectly correlated with the brightness of the comae of the impacting fragments: correlation was good for most fragments, but the fragments that were off the main line of nuclei tended to produce very minor impact effects.

Comet or asteroid?

Another important question is whether the nuclei in SL9 showed any signs of continuous dust and gas production (i.e. activity). This issue is intimately related to the question of whether SL9 was a comet as opposed to an asteroid, since asteroids generally do not show activity, while comets do. Viewed separately, each component of SL9 looked like a classical comet, complete with a nearly spherical coma and an elongated dust tail. However, the splitting of an asteroidal body would presumably release large amounts of dust that could also form comae and tails. (Once released, the dust particles move on orbits that are determined by their interaction with Jupiter and the Sun; whether the source of the dust was a comet or an asteroid is irrelevant.) Thus, we must search deeper before we can draw any firm conclusions regarding the nature of SL9.

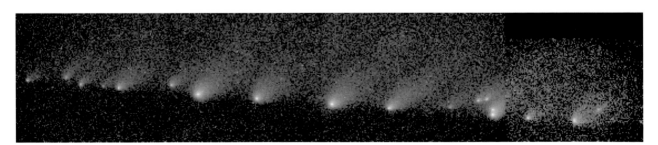

A Hubble Space Telescope image taken between 24 and 27 January 1994, the first of SL9 following the servicing mission. It was easily the best picture up to that time. All the nuclei can be seen except for W, which is to the right and out of the field of view. Several nuclei have drifted away from the main line of the train. Even at this high resolution it is not possible to distinguish the central nuclei from their enveloping dust clouds with any reliability. (H. A. Weaver & T. E. Smith, STScI/NASA)

A composite of two CCD images of SL9 obtained on 1 July 1994 with the Danish 1.5-m telescope at the ESO La Silla Observatory. It shows most of the individual fragments (A to S), just two weeks before the first impact. Fragment A is at the upper left. At the time of the exposure, the comet was about 11 million kilometres from Jupiter. (ESO)

The fact that the pieces of SL9 had symmetrical circular inner comae for most of its apparition argues for continuous dust production. Steady outflow of dust from each nucleus would naturally create a circularly symmetric coma. In contrast, an initially symmetric 'debris cloud' that was produced by dust released during the tidal break-up of the SL9 parent body and was not replenished by new dust, would presumably become elongated along the direction of the train, just as the train itself stretched out in that direction. Perhaps the fragments could have maintained circular debris clouds of old dust (i.e. dust released in July 1992) if the dust consisted of particles larger than about 1 cm (so that they are relatively unaffected by solar radiation pressure) moving outwards at very low speeds (5 cm/s

or less), but such an explanation seems rather ad hoc.

There is an important physical distinction between comets and asteroids: comets generally contain a significant fraction of their mass in the form of ice, while typical asteroids are essentially devoid of ice. In a classical comet, a small and solid nucleus is heated by the Sun and emits a stream of gas molecules that sublime directly into space from the ice. The escaping gas – mostly water molecules for comets around 1 AU from the Sun – then drags embedded dust grains from the nucleus and propels them into the coma where they are deflected out along the tail by solar radiation pressure. At 5 AU, SL9 was so far from the Sun that solar heat could barely sublimate water ice, even if it were fully exposed on the

A CCD image of fragments A to P taken in red light with the 3.5-m ESO New Technology Telescope on 11 July 1994. Fragment A was 5 million km from Jupiter at this time. Computer processing to reduce scattered light from Jupiter in the background produced the ring around the center of the picture. (ESO)

surfaces of the individual nuclei. On the other hand, it seems reasonable that a fractured nucleus might expose deeply buried ices to solar irradiation. For example, carbon monoxide (CO) and carbon dioxide (CO_2) are both known to be present in cometary nuclei, and in at least

one comet (P/Schwassmann–Wachmann 1, at 6 AU from the Sun) CO sublimation sustains a spectacular coma. In SL9, these 'supervolatile' ices would sublimate furiously and could produce the coma and tails observed in each component.

Following this argument, several groups of astronomers searched the spectrum of SL9 for emission lines characteristic of gases that would indicate cometary activity. On the ground, the Keck 10-m and Canada–France–Hawaii 3.6-m telescopes at Mauna Kea Observatory, Hawaii, and the McDonald Observatory 2.7-m telescope in Texas (and probably others) were used to search the visible part of the spectrum for lines due to CN, C_2, C_3, and other molecules, while the HST searched for OH, CS, and CO_2^+ in the ultraviolet. Submillimeter radio telescopes were used to search for spectral lines due to CO. Despite intensive efforts, no spectroscopic evidence for any of these gases was found. However, the upper limits deduced from the non-detections do not exclude the sublimation of gases as the source of the coma. Perhaps these null results can be explained if fresh supervolatiles exposed by fragmentation were lost in the year between the break-up and the spectroscopic observations.

Although spectroscopic observations of SL9 did not reveal any 'typical' cometary emissions, something very

Fragments L to W imaged through a red filter with the 3.5-m ESO New Technology Telescope on 15 July 1994. They were then so near Jupiter that scattered light from the planet was a severe problem. The attempt to subtract the bright background has produced some strange artifacts in the image. The brightest fragment is Q, which is close to the centre of the picture. (ESO)

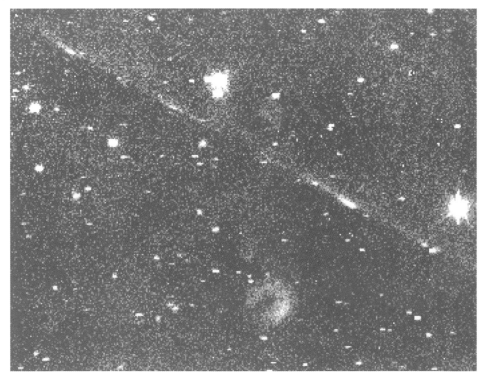

unusual happened to the spectrum during HST observations on 14 July 1994. While observing the G nucleus, the HST recorded a 2-minute outburst of emission from ionized magnesium followed 20 minutes later by a three-fold increase in the sunlight scattered by the dust near the wavelength 310 nm. During the outburst the color of the reflected sunlight also changed dramatically, brightening much more at longer wavelengths than at shorter wavelengths. These outbursts have tentatively been ascribed to small dust grains becoming electrically charged and then exploding as they passed from interplanetary space into the jovian magnetosphere. Since magnesium is an important constituent of both comets and asteroids, these intriguing observations probably do not shed any new light on the comet versus asteroid argument surrounding the nature of SL9.

In any event, the absence of gas emission lines certainly does not require that SL9 was an asteroid, since many short period comets at 5 and 6 AU from the Sun have extended dust comae but fail to show spectroscopic evidence for gas. On the other hand, even if the nuclei were shown to be ice rich, a cometary origin would still

not be assured, since it is likely that certain classes of so-called 'asteroids' are similarly ice rich. For example, the Trojan asteroids co-orbiting with Jupiter, are likely to contain substantial bulk ice, as are the more distant Centaurs and trans-neptunian objects. In this sense, the comet–asteroid dichotomy is something of a red herring, since the distinction between the two classes normally given becomes blurred for objects that formed in the trans-jovian region.

Another important clue regarding the nature of SL9 comes from tracking its orbit back in time. These calculations reveal a chaotic motion: within the very short time of 20 years the position of the comet becomes unknowable. But the key feature of the motion is that capture into an orbit like that of SL9 is very difficult if the starting orbit is highly eccentric like those of most long period comets. Orbits in which the initial velocity of encounter with Jupiter is low are much more likely to result in capture. The short period comets, which are already trapped around the Sun with the aid of Jupiter's gravity, are the most plausible source for such bodies. It is also possible, although less likely in a statistical

SL9 imaged with the ESO New Technology Telescope at the La Silla Observatory, Chile, on 1 July 1994. (R. Schulz & J. A. Stüwe, MPIA)

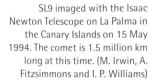

SL9 imaged with the Isaac Newton Telescope on La Palma in the Canary Islands on 15 May 1994. The comet is 1.5 million km long at this time. (M. Irwin, A. Fitzsimmons and I. P. Williams)

sense, that SL9 was originally a member of one of Jupiter's Trojan asteroid clouds. No physical observations tell us for certain what kind of object SL9 was, but the dynamical investigations hint at a cometary origin.

Though comet P/Shoemaker–Levy 9 is no more, scientists will continue to pore over the voluminous observations made of this fascinating object prior to its demise. The more we learn about the comet and its reaction to the extreme conditions to which it was subjected in the last years of its life, the better we will understand comets in general and the better we will be able to interpret the observations of the impacts themselves.

8

The Great Crash

John R. Spencer

The events of Impact Week, and its aftermath, were captured in images of Jupiter secured by observatories around the globe, in flight, in orbit and in space, which together covered almost the whole spectrum.

All our hopes and fears about the impacts would be resolved on July 16 1994, the date fragment A was due to strike Jupiter. Would this be the astronomical event of the century? Would we have to admit to the media and funding agencies that we had seen nothing whatsoever?

The preceding few weeks were very busy. There were last-minute predictions about what we would see, desperate attempts to get the latest state-of-the-art cameras and spectrographs up and running in time for the impacts, constant requests for information and interviews from the media, and travel plans to be made to unfamiliar destinations. We also tried to calculate in advance every parameter that we could think of that would be useful for our observations. I produced and distributed a table that showed how to find the locations of the impact sites for about three days after each collision, using the Galilean satellites as a reference. I figured that we would need a way to find the impact sites, which would presumably be invisible in our images, and that any effects of the impacts would be unlikely to last longer than three days. I was wrong on both counts. On 14 July Paul Weissman of JPL had an article published in the journal *Nature* entitled 'The Big Fizzle is Coming'. But for all the public pessimism, it was an exciting time. Perhaps astronomers weren't really as pessimistic as they claimed to be publicly.

A great exodus of astronomers ocurred in the week before the impacts. Around the world, offices emptied and observatory dormitories filled up. The average direction of migration was southward. Most astronomers live and work in the northern hemisphere, but Jupiter was in the southern skies and best seen from the southern hemisphere. Nevertheless, every major observatory in the world, in both hemispheres, would be watching the impacts. Each impact could be seen only from the side of Earth that happened to be facing Jupiter at the time. With 20 impacts to choose from, everyone would get a chance to observe several of them, but the turning Earth handed the crucial first event to Europe, Africa and South America. Antarctica would see all the events, including the first.

Impact Week

The impact of fragment A was predicted to take place at 20:00 UT on 16 July, give or take about 15 minutes. The German/Spanish team at the Calar Alto Observatory in Southern Spain, led by Canadian Tom Herbst, saw nothing unusual on Jupiter till 20:11 UT, when they suffered a computer crash that halted observations. After four nerve-wracking minutes they were back in business at 20:15 UT. Jupiter still showed nothing unusual (though subsequent processing revealed that a very faint dot had already appeared on Jupiter's limb just before the computer crash). They continued to take images as rapidly as possible.

Then, at 20:17:30 UT, Tom noticed a small dot of light on the limb of Jupiter, right where the impact was supposed to take place. It was happening! The dot soon began to brighten rapidly. Incredibly, it became as bright as Io in matter of a few minutes before fading again. A stunned silence fell over the control room. Never before had astronomers seen such a brilliant and rapid intrinsic change in an astronomical object. At the same moment, astronomers in the Canary Islands, in South Africa, at the South Pole, and in Chile were transfixed by the same sight. The Calar Alto team rushed out a deceptively sober message via the SL9 e-mail Exploder, becoming the first to announce to the world that impact week would disappoint no-one.

> "Impact A was observed with the 3.5-m telescope at Calar Alto using the MAGIC camera. The plume appeared at about nominal position over the limb at around 20:18 UT. It was observed in the 2.3 micron methane band filter brighter than Io."
>
> *Tom Herbst, Doug Hamilton, Jose Ortiz,*
> *Hermann Boehnhardt, Karlheinz Mantel and Alex Fiedler*
> *at Calar Alto Observatory, Spain, 16 July 1994, 21:28:15 UT*

After the flash of the first impact faded, it left something behind – a huge cloud that appeared as a bright spot in methane-band images. This persisted as Jupiter's rotation carried it away from the limb and across the disk of the planet. When amateur astronomers in England and Florida saw this persistent cloud as a small but distinct dark spot at visible wavelengths, it became clear that impact week would be a spectacle for everyone to appreciate, not just the professionals.

The Hubble Space Telescope had also been lucky enough to be on the right side of the Earth in its orbit when the A impact occurred. Not knowing the exact time of the impact in advance, the HST had been programmed to take a continuous series of images at a set of carefully-chosen wavelengths in the hopes that some of them would show something. These images were stored on the onboard tape recorder for about four hours, until a communications satellite was available to transmit them to Earth. By the time the images appeared on the monitors at the Space Telescope Science Institute in Baltimore, the waiting crowd of scientists had already heard the initial Exploder reports from Spain and Chile, but they were still astonished and exhilarated by what the HST's superior

vision revealed. A great plume, captured perfectly in profile on Jupiter's limb, rose thousands of kilometers above the cloudtops before collapsing.

On the HST's next orbit of Earth, when the impact site had had time to rotate into view, the telescope obtained a detailed view of the black cloud that amateur astronomers had seen, revealing a beautifully symmetrical scar with a central core and semicircular halo around it. Champagne bottles were opened, and camera team leader Heidi Hammel grabbed crude printouts of the images and dashed upstairs to the auditorium where Gene and Carolyn Shoemaker, unaware that the images had arrived, were addressing a press conference. The Shoemakers did not mind when Heidi interrupted them with the news. Impact week was in full swing.

We will let the pictures tell most of the rest of the story of that remarkable week. The A impact was followed by others of similar brightness as the rest of the inital group of small fragments, A–F, slammed into the planet over the next 30 hours. Strangely, fragments B and F produced almost no effects at all, though some observatories reported faint flashes or hints of faint clouds at their impact sites. These fragments looked similar to their neighbors prior to impact, with one exception. Most fragments fell along a straight line in the pre-impact images, but B and F, (and P, T, U and V, which also 'fizzled' on impact) were displaced from the line in a direction away from the Sun. Perhaps they were so flimsy that evaporating gas on the sunlit side had been able to push them back off the line.

The impact of the four bright fragments in the central part of the comet train, G, H, K and L, produced the most dramatic effects of all. Impact fireballs briefly outshone the planet itself at certain wavelengths and the huge black spots created on Jupiter stunned observers worldwide. Impact L may have been the largest of all, but G and K were close, with H being somewhat fainter. The K event was notable for what was not seen; Jupiter's large satellite Europa was lurking, invisible, in Jupiter's shadow at the time of the impact, providing a perfect reflector for the brilliant but hidden flash that many expected at the moment of impact. Europa did not light up in the dark enough to become visible, however. Despite the brilliant infrared flashes, the impacts did not appear to be very bright at visible wavelengths.

The 'Big Four' were followed by a mixed bag. The N impact, which was faint but distinct, the P2 and P1 impacts, which were not definitively detected, the Q2 event, which was very faint and left a barely-visible scar in

The comet heads into the glare of Jupiter in its final days. This remarkable photographic image is one of the few that includes both Jupiter and the comet in the same frame, though Jupiter (over 100 million times brighter than the comet) is enormously overexposed and SL9 is only faintly visible, at about '4 o'clock' relative to Jupiter, as a line of nuclei and dust pointing directly at the planet. It was taken on 12 July 1994 by Guido Pizzaro with the ESO 1-m Schmidt telescope. (ESO; photographic enhancement Hans Hermann Heyer)

HST images, and the bright Q1 impact. Q had appeared to be the brightest fragment in 1993, but in the end its impact was fainter than G, K or L's. By this time it was becoming difficult to detect faint impacts and impact sites simply because Jupiter's southerly latitudes were becoming very crowded with the debris of previous collisions.

The week wound down with impact R, quite bright and well-observed, impact S, also bright, the very faint T, U and V impacts, and the final W impact, which closed the proceedings with a respectable bang, and was observed by the HST and *Galileo* as well as observatories on Earth.

The end of the impacts did not mean the end of observations, of course. By the end of the week Jupiter was a scarred and disturbed planet, pushed far from equilibrium by the pounding it had received, its appearance anomalous at wavelengths from the far ultraviolet to the radio.

It would be a fascinating task to watch the planet gradually return to normal, but this would not require the continuous coverage that impact week had demanded, and the world's telescopes were needed for other duties. So most of us copied our data onto magnetic tapes for safekeeping and headed home.

The Lowell Observatory portable observatory, shipped out from Arizona, is set up in the outback for Jupiter impact monitoring at Charters Towers, NE Australia. Several similar systems were deployed around the world in regions remote from large telescopes. A computer-controlled CCD camera is attached to the Celestron 14-inch telescope. Marc Buie (shown here) and Larry Wasserman took 50 000 images of Jupiter with this system during the impact period. (Larry Wasserman & Marc Buie)

The Calar Alto 3.5-m telescope, the largest optical telescope in Europe, operated by the Max-Planck-Institut für Astronomie in Heidelberg, Germany, and the Centro Astronomico Hispano-Aleman in Almeria, Spain. A team from the Max Planck Institut, led by Tom Herbst, mounted an infrared camera called 'MAGIC' on this telescope to obtain images of the comet impacts. (Courtesy Tom Herbst)

The first dramatic report of the effects of the impact of fragment A came from Calar Alto observatory in southern Spain, which produced some of the finest images throughout the impact week. This aerial view of the observatory shows the 3.5-m telescope dome in the foreground. (Courtesy Tom Herbst)

The Calar Alto 3.5 m telescope observing team watch the impact images on their computer screen in the telescope control room, where they could view images as soon as they were taken. From left to right, Hermann Boehnhardt, Jose Luis Ortiz, Doug Hamilton, Tom Herbst (at the computer) and Karl-Heinz Mantel. (Courtesy Tom Herbst)

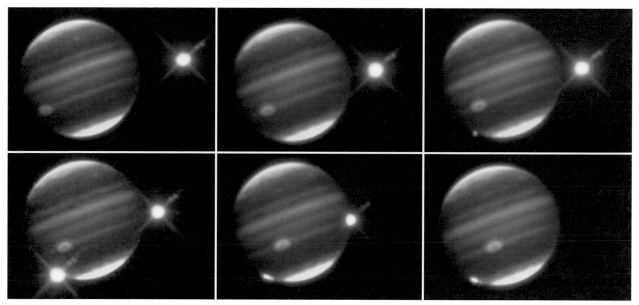

The A impact seen with the MAGIC infrared camera on the 3.5-m telescope at Calar Alto Observatory at a wavelength of 2.3 microns, where methane absorption darkens Jupiter dramatically and renders the impacts more visible. In all but the last frame, where it has passed behind Jupiter, the moon Io is brightly visible on the right side of the images, and the Great Red Spot is conspicuous as a pale oval on Jupiter which appears to move across the disk as Jupiter rotates.

The first image, taken at 19:49 UT, 22 minutes before impact, shows Jupiter looking as it had always looked at this wavelength. The second image, taken close to the moment of impact, shows the first hint that impact week would be spectacular: a faint brightening on the edge of Jupiter just below the Great Red Spot. In the third frame, five minutes later, the spot is beginning to brighten. Tom Herbst and his colleagues were forced to shorten the exposure time to avoid overexposing the image. In the fourth frame, 12 minutes after impact, the explosion has reached maximum brightness, brighter than Io and far brighter than anyone had anticipated. In the fifth frame, taken 31 minutes after impact, the impact spot has faded but has grown as large as Earth. In the sixth frame, 38 minutes after impact, the impact site has rotated further into direct view. A bright 'core' and fainter 'halo' to the south can be seen, setting a pattern that most subsequent impacts would follow. (T. Herbst, Max-Planck-Institut für Astronomie, D. Hamilton, Max-Planck-Institut für Kernphysik, H. Boehnhardt, Universitäts-Sternewarte, München, & J. L. Ortiz, Instituto de Astrofisica de Andalucia)

The A impact portrayed in a sequence of infrared images (2.2 micron band) taken with the 0.75-m telescope of the South African Astronomical Observatory. The sequence starts at the lower right at 20:17 UT and ends at the top left at 20:32 UT. the frames are spaced at one-minute intervals. Io is the bright spot to the right of Jupiter. (K. Sekiguchi, SAAO)

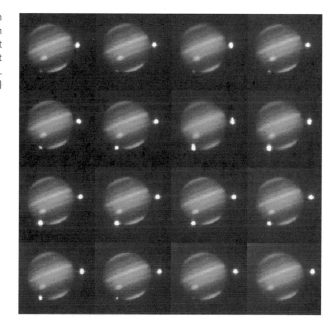

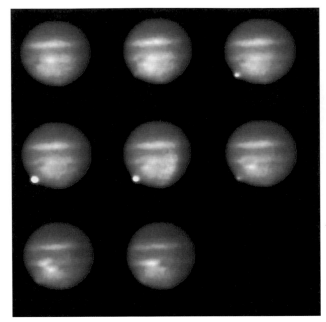

The A impact was also spectacular at a wavelength of 12 microns, as seen in this sequence from the CAMIRAS long-wavelength infrared camera at the 2.56-m Nordic Optical Telescope (NOT) on the island of La Palma in the Canary Islands. It covers the interval 20:12 to 21:13 UT, from top right to lower center. Jupiter is brighter at this wavelength than at 2.3 microns, so the impacts appear relatively fainter. A glow at the impact site remains faintly visible in the final image, one hour after impact. (P.-O. Lagage, P. Galdemard, R. Jouan, P. Masse, E. Pantin, J. A. Belmonte, G. Olofsson, M. Hultgren, J.-A. Belmonte, A. Ulla)

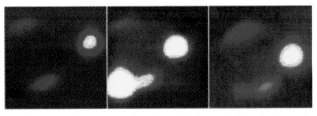

SPIREX images taken at a wavelength of 2.3 microns before, during, and after the A impact (between 20:00 and 21:00 UT on 16 July). Io is visible on the right in all three frames. Left frame: Jupiter's north and south polar haze caps mark the position of the planet on the left side of the frame. Center: The A impact can be seen as a very bright spot. Right: A faint remnant of the impact is still visible. Because SPIREX is a small telescope and Jupiter was only 12 degrees above the horizon, this and other SPIREX images have low resolution, but they have the unique advantage that continuous monitoring of Jupiter was possible during the six-month polar night. The observers were University of Chicago astronomers Mark Hereld, Hien Nguyen (at the South Pole), Bernard J. Rauscher, and Scott A. Severson. (By special permission of the Center for Astrophysical Research in Antarctica (CARA) and the University of Chicago. Copyright CARA and the University of Chicago)

(Right) Astronomers at the Space Telescope Science Institute in Baltimore crowd around a computer monitor to get their first glimpse of the impact of fragment A as observed by the HST. The observations were transmitted from the telescope to the control room several hours after the actual impact. (STScI)

SPIREX (South Pole Infrared Explorer) observers Mark Hereld, Bernard Rauscher and Hien Nguyen at the south pole. This image was taken with a digital camera, transmitted via satellite to Chicago, and distributed on the Internet during impact week. (By special permission of the Center for Astrophysical Research in Antarctica (CARA) and the University of Chicago. Copyright CARA and the University of Chicago)

This series of seven images was obtained with the TIMMI far-infrared instrument at the ESO 3.6-m telescopes on 16 July, between 20:23 UT and 21:26 UT. It shows the plume rising above the site of the impact of fragment A. The wavelength range covered was 9.4–11 microns. (B. Mosser, T. Livengood & U. Kaeufl/ESO).

The scar left by fragment A as seen by the HST one and a half hours after impact. The impact site is marked by a broad dark streak surrounded by a partial ring, several thousand kilometers across. The image was taken in violet light (410 nm) with the WFPC2. The comet fragment entered Jupiter's atmosphere from the south in the direction of the streak at an angle of about 45° from the vertical. (H. Hammel, MIT/NASA)

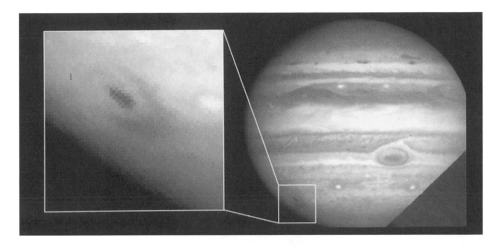

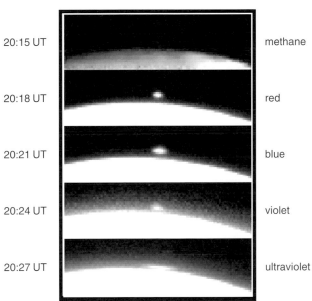

20:15 UT	methane
20:18 UT	red
20:21 UT	blue
20:24 UT	violet
20:27 UT	ultraviolet

This sequence of HST images shows a plume created at the time of the A impact. It can be seen rising 1000–1500 km above the limb of Jupiter at 20:18 UT in the red image. The first image, at 20:15, does not show any such feature. Later images clearly show the plume spreading. It was visible at wavelengths ranging from the ultraviolet through the near infrared (HST Jupiter Imaging Science Team)

At Cerro Tololo Inter-American Observatory in Chile, observers were frustrated by clouds throughout impact week, and missed the A impact even though it was seen from the European Southern Observatory only 100 km to the north. Beautiful sunsets, here illuminating the dome of the 4-m telescope, were small consolation, but the clouds parted in time to allow some of the first observations of the persistent cloud at the A impact site. (J. Spencer)

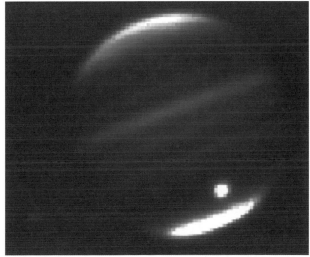

The brilliance of the A impact site in the 2.3 micron methane band three hours after impact shocked observers at the Cerro Tololo 4-m telescope when they saw their first images. The impact produced clouds very high in the jovian atmosphere, which reflected sunlight before it could be absorbed by the methane at lower levels that darkens the rest of the disk. (John Spencer & Darren Depoy)

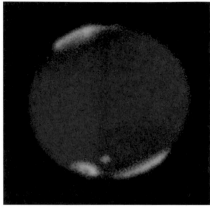

Some of the first reported observations of the impact sites at visible wavelengths were made by amateur astronomers in Florida. Don Parker was setting up his 40-cm (16-inch) telescope in the afternoon of 16 July when he got an excited phone call from Jeff Beish, who lived nearby. Broad daylight notwithstanding, Jeff had just seen a black spot on Jupiter at the A impact site. They reported this observation to Brian Marsden at the Central Bureau for Astronomical Telegrams and, just as the Sun was setting, Don Parker was able to obtain the CCD image on the right at 00:11 UT on 17 July. The image on the left show the planet on 14 July, minus the impact scar. South is at the top in these images, following the convention used by amateur astronomers. (Don Parker)

The C impact revealed new surprises at wavelengths between 3 and 4 microns. This image, taken with the CASPIR camera on the Australian National University's 2.3-m telescope at Mount Stromlo, shows a large ring spreading out from the impact site after it has rotated into direct view. The A impact site is also visible, without a surrounding ring, and Jupiter's aurorae glow orange at each pole in this color rendition. (Peter McGregor, ANU)

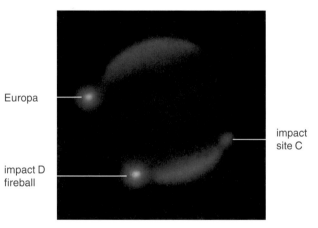

Just as impact site C was about to rotate out of view after its first transit of Jupiter's disk, fragment D hit. It was similar in brightness to impacts A and C. This image at 2.34 microns is from the CASPIR camera on the Australian National University's 2.3-m telescope at Siding Spring Observatory. (Peter McGregor, ANU)

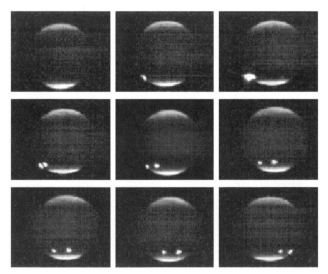

Unlike the B impact, the C impact on 17 July did not disappoint. These images at 2.37 microns from the Anglo-Australian Telescope show, in sequence from the upper left: 6:10 UT, Jupiter's normal appearance at this wavelength; 6:45 UT, the reappearance of the A site as it completed its first jovian rotation; 7:13 UT, the flash of the C impact, appearing near the A site due to foreshortening; 7:29 UT, the fading and spreading of the C flash; 8:05, 8:53, 9:07, 9:35 and 10:49 UT, the rotation of the two sites across Jupiter. (D. Crisp, V. Meadows, S. Lumsden & S. Lee. Copyright Anglo-Australian Telescope Board)

> "The SPIREX fragment D data was significantly compromised due to the sudden onset of low blowing snow. The telescope was heroically cleared of accumulated snow by Joe Spang of the AMANDA project, and John Briggs of the ATP project, in strong winds at temperatures of −60 degrees Celsius."
>
> *Mark Hereld, Hien Nguyen, Bernard J. Rauscher, Scott A. Severson of the Astronomy & Astrophysics Center, University of Chicago, at the South Pole, 18 July 1994, 01:44 UT*

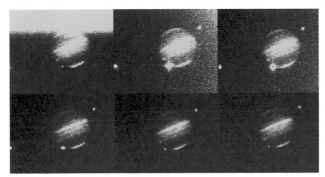

The plume produced by the impact of fragment E in a sequence of images at 2.3 microns from the 0.75-m telescope of the South African Astronomical Observatory. The sequence starts at 15:16 UT on 17 July and runs from lower right to upper left at two-minute intervals, except between the first two frames where the interval is 5.5 minutes. The bar across the first frame and background unevenness results from the problem of observing in daylight. (K. Sekiguchi, SAAO)

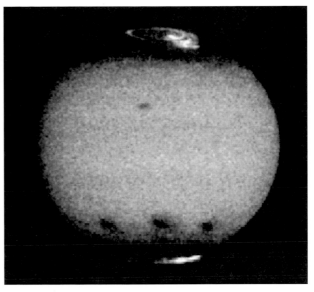

This far ultraviolet (160 nm) image taken at about 19:00 UT on 17 July with the WFPC2 on the Hubble Space Telescope shows as dark marks the sites where fragments C, A and E (in order from left to right on the disk) had struck Jupiter about 12, 23 and 4 hours earlier, respectively. The bright areas near the poles are aurorae and Io is crossing above the center of the disk. (John Clarke, University of Michigan/NASA)

The large bright impact site of fragment E (left) and the fainter remnant of the D impact (right) are seen in this image from the Cerro Tololo 4-m telescope made by combining three infrared wavelengths: 1.58 microns (blue, no methane absorption), 1.70 microns (green, weak methane absorption) and 2.30 microns (red, strong methane absorption). This color combination shows the height of the clouds and haze in the atmosphere. The deepest clouds appear blue, mid-level hazes (such as above the Great Red Spot) blue-green, high haze layers (such as the polar hazes) orange, and the highest hazes (at the impact sites) red. The impact clouds seemed to be the highest ever seen in Jupiter's atmosphere.

This image was taken at 02:15 UT on 18 July, less than two hours after the predicted impact time of fragment F, which hit close to the E site but produced only a very faint flash. At the time of this image a vigorous global discussion was in progress among observers about the identity of the bright cloud. Was it E or F, or a combination? We are now fairly sure that the cloud is mostly from the older E impact, but such confusion only got worse as the week progressed and overlapping impact sites multiplied. (John Spencer, Darren Depoy & Nick Schneider)

A visible-wavelength image of Jupiter taken with the 3-m Shane telescope at Lick Observatory on 18 July at 04:29 UT. It has been computer processed using a speckle image reconstruction technique. This view is one jovian rotation later than the HST far UV image above. Dark spots are visible at the impact sites of fragments C, A, and E. (C. Max, D. Gavel, E. Johansson, M. Liu, W. Bradford/Lick Observatory/ Lawrence Livermore National Laboratory)

There was eager anticipation of the impact of the G fragment, which appeared brighter than all previous fragments in pre-impact images. Observers were not disappointed: G may have been the biggest impact of all. The fireball is seen here in a 2.34-micron infrared image made with the CASPIR camera on the ANU's 2.3-m telescope at Siding Spring Observatory. The remains of the C impact site is faintly visible on the opposite side of Jupiter's disk. (Peter McGregor, ANU)

“Looking through fog and 98% humidity, we imaged impact G at 2.29 and 4.78 microns... The series began at 7:35 UT, when precipitation stopped briefly, with the site already bright. The impact region increased sharply in brightness at 4.78 microns, eventually saturating the detector during its brightest phase. Although the images did not contain the entire planet, we estimate that the impact site outshone the planet at 4.78 microns. Rapidly varying fog conditions prevented a real-time analysis of the 2.29-micron images.”

Orton, Baines, Esterle, Friedson, Goguen, Harrington, Kaminski, Lisse, Miller, Shure (NASA IRTF Comet Collision Team) on Mauna Kea, Hawaii, 18 July 1994, 08:47 UT

The 10-m Keck telescope on Mauna Kea, by far the world's largest optical telescope. Plagued by damp conditions, the observers there opened their dome when they heard that their neighbours at the IRTF were doing so, and glimpsed the brilliance of the G impact before weather conditions forced them to close the dome again. This time exposure image of the huge Keck dome was taken during more favorable conditions. (Russ Underwood, for the California Association for Research in Astronomy)

“The summit here in Hawaii is plagued by heavy fog. All telescopes are closed. At 21:27 local time, a minute or so before the expected impact, the IRTF noticed a clearing, and we opened up. At 21:39 we obtained our first frame of Jupiter, in regular K-band: a truly remarkable (saturated) plume was visible well above the limb. We started a sequence of observations at 3.4 microns: a spot was visible in our first frame at 21:40, which brightened to truly remarkable levels by 21:50, after which it decreased in intensity. At the same time the fog was coming in, and we were closing up again; whether we were seeing a true decrease in the spot's intensity, or whether the decrease is due to increasing cloud coverage above the telescope is not yet known. It is raining right now; we don't expect to get any more frames tonight.”

Imke de Pater, James Graham, Garrett Jernigan and collaborators at the W. M. Keck Observatory, Hawaii, 18 July 1994, 09:51 UT

The G impact was observed by the Kuiper Airborne Observatory (KAO), flying out of Melbourne, Australia. The KAO, a converted C-141 cargo plane, uses its 0.9-m telescope to make astronomical observations from an altitude of 12 500 m (41 000 feet) where it is above 85% of the Earth's atmosphere and 99% of the water vapor. This allows it to observe radiation at wavelengths that are absorbed by water vapor and are thus undetectable from the ground. One of the hopes of the astronomers who used it during impact week was that they might detect water vapor in the impacts, without confusion from water vapor in our own atmosphere. This view of the KAO in flight shows the square hole just forward of the wings through which the telescope makes its observations. (Courtesy NASA Ames Research Center)

There was jubilation aboard the Kuiper Airborne Observatory when the team detected strong 7.7-micron methane emission lines from the G impact. Gordy Bjoraker (extreme left) celebrates with the astronomers whose KEGS spectrograph he is using for this project: Terry Herter (center, standing), Susan Stolovy (center, seated), and George Gull (standing in the background, in front of the circular telescope housing), and with members of the KAO flight crew. Everyone on the plane uses headsets to communicate since the cabin is very noisy. Bjoraker and his team later realised that the spectra they had obtained showed not just methane, but also water vapor, and water vapor was also seen by the astronomers who observed some of the later impacts with the KAO. (Courtesy Richard Cisar-Wright, *The Australian* newspaper)

The G impact produced the most dramatic impact scar yet seen, as shown in this HST image. It was taken between 09:19 and 09:25 UT, one and three-quarter hours after impact, when the site had rotated onto the visible disk. A dark ring, presumably some type of wave, spreads outwards at 2000 km/hour from the impact site, which is at one end of a dark streak that marks the approximate direction of the comet's entry into the atmosphere. Beyond the circular wave is a huge (Earth-sized) asymmetric halo. The much smaller site of the D impact, now 28 hours old, can be seen to the left of the E site. (Heidi Hammel and the HST Comet Team; additional processing by Robin Evans (JPL); courtesy John Trauger)

Impact H was also a big one, though reported to be fainter than G by the South Polar team (the only people able to see both). This infrared methane-band sequence was taken with the 3.5-m telescope at Calar Alto Observatory, starting at 19:28 UT on 18 July. The first frame shows Jupiter's appearance before the H impact: the 12-hour-old G site, with its core and halo, is conspicuous and the faint D site can just be seen immediately to its left. Ganymede, in the upper left, is about to pass in front of Jupiter.

The second frame, taken about two minutes after impact, shows a faint precursor flash, which was followed four minutes later by the much brighter main flash, seen approaching peak brightness in the third frame. The observing team took spectra of the impact flash during peak brightness, so there are no images at that time. The fourth frame, 23 minutes after impact, shows the new impact site rotating into view, still fading but already huge. The fifth image, 25 minutes later, shows the fainter halo to the south of the bright core as the site takes on the now-familiar halo/core structure of the larger impact sites. Finally, the sixth image, taken over three hours after impact, shows four major impact sites (from left to right, A, E, H and G). The fresh H site has now rotated past the central meridian, and the Great Red Spot has come into view. Jupiter is beginning to look seriously battered. (T. Herbst, D. Hamilton, H. Boehnhardt & J.- L. Ortiz)

The first view of the day-old G impact site from North America, on the evening of Monday 18 July local time, stunned professionals and amateurs alike. In nearly 400 years of observations, no-one had ever seen Jupiter look like this before. This sequence was taken between 03:32 and 06:52 UT on 19 July at Lick Observatory with the 3.05-m telescope using a 'speckle' camera system, which combines many short exposure images to compensate partially for the blurring caused by Earth's atmosphere. It shows the G site rotating across Jupiter's disk, with the H site just appearing in the final frame. Images such as this do not really do justice to the visual impact of the view seen directly through even a small telescope. (C. Max, D. Gavel, E. Johansson, M. Liu, W. Bradford/Lick Observatory/ Lawrence Livermore National Laboratory)

> "I want to put this into the historical context of Jupiter observations. It is now about 5:30 UT, 19 July. I have just come in from looking at Jupiter with my back yard telescope. (Our team will again be on the Kitt Peak 2.1-m telescope three evenings from now.) The preceding end of impact site G is approximately on the central meridian. Based on my own extensive experience of observing Jupiter when I was younger, and studying historical records of Jupiter observations from the early drawings of Hooke and Cassini through the extensive 19th and 20th century reports of the British Astronomical Association, I would assert: THIS IS THE MOST VISUALLY PROMINENT DISCRETE SPOT EVER OBSERVED ON JUPITER. (By 'prominence' I mean the combination of both size and contrasting albedo.) Does anyone disagree?"
>
> *Clark R. Chapman, Arizona, 19 July 1994, 05:40 UT*

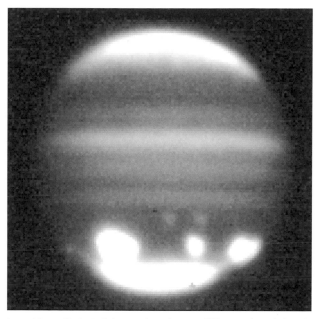

A chain of five impact sites glow brightly in this infrared image (1.7 microns) taken with the MAGIC camera on the 3.5-m telescope at Calar Alto Observatory at about 21:00 UT on 19 July. From left to right they are K, C, A, E and H. The Great Red Spot is just disappearing on the right while two white ovals, persistent features in the jovian atmosphere are rotating into view on the left, north of the impact sites. (T. Herbst, K. Birkle, U. Thiele, D. Hamilton, H. Boenhardt, A. Fiedler, K.-H. Mantel, J.-L. Ortiz, G. Calamai & A. Richichi)

Impact L was the last of the very large impacts. This 2.3-micron infrared image taken at 22:14 UT on 19 July with the Calar Alto 3.5-m telescope shows a mystery associated with it: a very faint spot appeared on the limb at least 20 minutes before the L impact, and lingered there until the main impact overwhelmed it. Presumably some small and previously unseen piece of the comet preceded L to destruction and left this brief record of its existence. The location of the spot is approximately consistent with its being the 20-hour-old impact site of the J fragment, which disappeared in 1993, but no-one else reported seeing the J impact site. Also visible here are the overexposed images of sites K, C, A and E: the images of A and E are merged. (T. Herbst, D. Hamilton, H. Boehnhardt & J.-L. Ortiz)

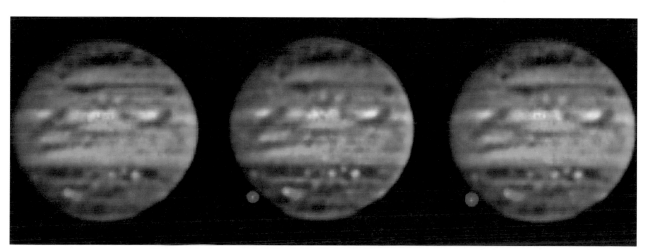

Impact L, captured in one of the few '3-color' infrared sequences of an impact flash. They were taken with the 1-m telescope at the Pic du Midi Observatory at 22:18, 22:20 and 22:23 UT on 19 July. The earliest image, (on the left) was taken one minute after impact and shows the very faint 'precursor' flash in the original. The center and right images show the much brighter main flash, which appears red because it is brightest at longer wavelengths: 2.2-micron radiation is represented by red, 1.6 microns by green and 1.2 microns by blue. The sites of the K, C and A impacts can also be seen as red spots. (F. Colas (Bureau des Longitudes), J. Lecacheux (Observatoire de Meudon), P. Laques (Observatoire Midi Pyrénnées)/ Station de Planetologie des Pyrénnées)

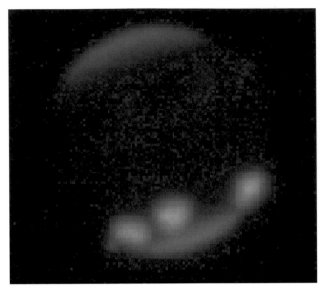

The impact of fragment N was one of the faintest visible. On this 2.34-micron image taken at 10:37 UT on 20 July with the ANU's 2.3-m telescope at Siding Spring Observatory, the faint plume is just discernible at the lower left edge of Jupiter's disk but several previous impact sites are much more prominent. (Peter McGregor and Mark Allen/ANU)

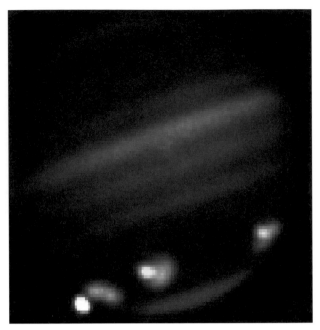

The Q impact was initially expected to be the brightest, but it was actually fainter than G, K, L and probably H. However, four separate impacts were seen associated with the Q family of fragments. Three were bright, with two fainter ones interspersed. This 1.7-micron infrared image from the Calar Alto 3.5-m telescope shows the Q1 fireball near maximum brightness, with previous impact sites visible further to the right. From left to right they are G and D (merged), L near the center and K near the right limb. (T. Herbst, K. Birkle, U. Thiele, D. Hamilton, H. Boenhardt, A. Fiedler, K.-H. Mantel, J.-L. Ortiz, G. Calamai & A. Richichi)

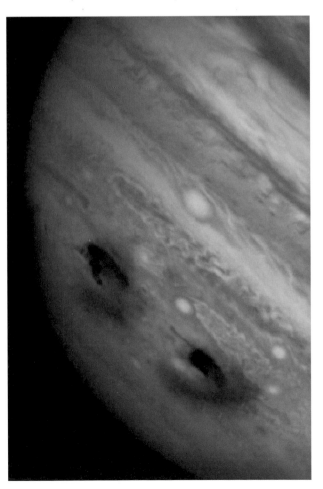

By the time of the Q impact the black clouds at the nearby G site (left) and L site (right) were beginning to show their age as the jovian stratospheric winds started to pull them apart. In this HST view, taken between 20:53 and 20:59 UT on 20 July, the core of the 2.5-day-old G site is breaking into many dark spots and its northern edge has been stretched out by east-blowing winds, and the halo is also becoming distorted. Similar processes are starting to act on the 0.9-day-old L site. The halo of the fresh Q1 impact site is beginning to rotate into view on the terminator, and a tiny dot between it and the G site marks the Q2 impact. (Courtesy H. Hammel & HST Comet Team/NASA)

This image in light of 700 nm wavelength of the G and L sites was taken at 21:11 UT on 20 July, just after the Q impact, at Pic du Midi, which has a reputation for having some of the best seeing anywhere on Earth's surface when conditions are good. The sharpness of this image compares quite favorably with the the near-simultaneous HST image. It also illustrates the dramatic appearance of this face of Jupiter: the planet appeared to have two immense eyes, formed by the G and L impacts. (F. Colas (Bureau des Longitudes), J. Lecacheux (Observatoire de Meudon), P. Laques (Observatoire Midi Pyrénnées)/ Station de Planetologie des Pyrénnées)

At least eight impact sites are visible simultaneously in this HST image, taken at 09:29 UT on 21 July, four hours after the R impact. They are, from left to right: E (on the terminator), H (with a strange four-leaf clover appearance), N (a tiny dot), Q1, Q2 (another tiny dot), the fresh R site, D (hard to distinguish from G) and G (very large, near the limb). The Great Red Spot is also prominent. (H. Hammel and the HST Comet Team; additional processing by Robin Evans (JPL); courtesy John Trauger)

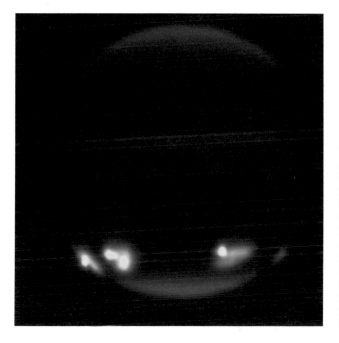

The R impact was one of the best observed, due to good weather over North America, the Pacific, and Australia. This 2.3-micron image of the impact flash taken at 05:37 UT on 21 July with the University of Hawaii's 2.2-m telescope on Mauna Kea is one of the highest resolution ground-based images of the impacts. It owes its detail to two technological innovations. First, an adaptive optics system followed the motion of the image of Io (off the frame to the left) as it danced around due to atmospheric turbulence, and removed this motion by rapidly moving a mirror to compensate. Second, the observers used one of the first large-format infrared cameras, which has an array composed of 1024×1024 pixels rather than the 256×256 pixels found in most infrared cameras. From left to right, the R flash appearing behind the G site, then the L, K and C sites. (K. Hodapp, J. Hora, K. Jim, & D. Jewitt, processed by R. Wainscoat and L. Cowie/University of Hawaii)

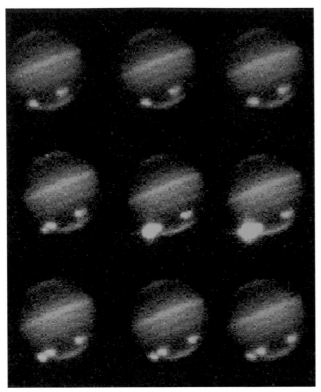

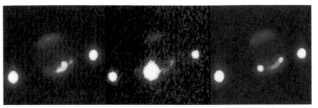

Impacts T, U and V were very faint, but the W impact closed Impact Week with a flourish. It is seen here at 08:14 UT on 22 July in 2.36-micron images from the 0.6-m SPIREX telescope at the South Pole, which with its privileged location observed the first and last impacts and at least nine others inbetween, probably more than any other observatory. The central image is noisy because a passing cloud temporarily obscured the view. Io is visible to the left of Jupiter, with Ganymede to the right, and the E and H impact sites can be seen on Jupiter's disk. (M. Hereld, H. Nguyen, B. Rauscher, S. Severson & J. Lloyd)

A series of images showing impact S in the 2.3-micron infrared wavelength band. They were taken with the 0.75-m telescope at the South African Astronomical Observatory at 5-minute intervals between 15:09 and 15:55 UT. (K. Sekiguchi, SAAO)

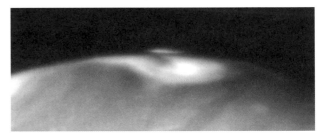

The HST also caught the final impact at 08:20 UT on 22 July. This image, taken in the 889-nm methane band, shows the collapsing W plume illuminated by morning sunlight 14 minutes after impact. In the foreground is the K impact site, appearing bright because sunlight is reflected before it can absorbed by the methane deeper in the atmosphere. The K site is now nearly 3 days old and has been severely distorted by the stratospheric winds. (H. Hammel and the HST Comet Impact Team/NASA; additional processing J. Spencer)

Frank and Ernest

Reprinted with the permission of Bob Thaves.

News of the impacts spread far beyond the world of astronomy. Reports appeared on the front pages of most newspapers, on national radio and TV news bulletins, and in news magazines worldwide. It was obvious that the comet crash had thoroughly worked its way into the public consciousness when the story spread from the front page to the comic pages. This cartoon appeared in US newspapers on 20 August 1994. (Reprinted with the permission of Bob Thaves)

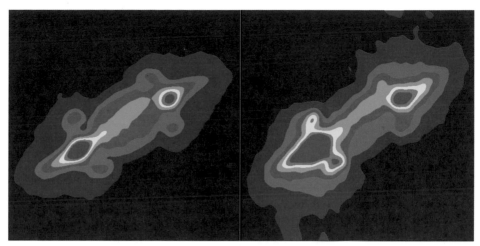

There were many predictions of the possible effects of the impact on near-Jupiter space, ranging from brightening of the jovian ring to a temporary shutdown of the Io plasma torus, but few of these predictions came to pass. One near-Jupiter effect that was clearly seen, however, was a dramatic brightening of the synchrotron radio emission that comes from electrons trapped in Jupiter's magnetic field. The electrons form a complex torus tightly wrapped around Jupiter, and this can be mapped at radio wavelengths. These two images, made at a wavelength of 20 cm with the Very Large Array in New Mexico, show, on the left, the electron torus on 24 June 1994, before the impacts, and on the right, the same side of Jupiter on 19 July, during the course of the impacts. They clearly demonstrate the increase in brightness during impact week. Jupiter itself is almost invisible at these wavelengths. Its disk occupies most of the space between the bright peaks in emission (red). The cause of the increase is not yet understood. Perhaps the incoming comets, or the impacts themselves, released large numbers of electrons into the magnetosphere. (Imke de Pater, University of California, Berkeley/VLA/NRAO)

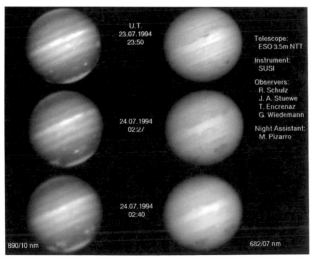

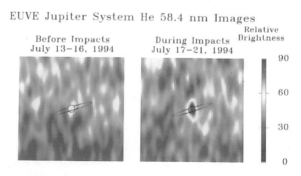

In the days and weeks that followed the impacts, astronomers continued to keep a close watch on Jupiter, wondering how the impact sites would change and how long they would last. This sequence of CCD frames, taken with the ESO 3.5-m New Technology Telescope at La Silla in Chile, shows Jupiter's appearance in red light (682 nm, right) and in the 890 nm methane band (left) on 23–24 July, two days after the impacts ended. The impact sites continued to be dark in the visible and bright in the methane bands. The top view (23:50 UT, 23 July) shows the A site on the central meridian and the more conspicuous E site to its right, and the H site disappearing over the right limb. The second pair (02:27 UT, 24 July) shows, from left to right, the L and K sites and, in the third image pair, taken 13 minutes later, the G site is rising on the morning terminator. (R. Schulz, J. A. Stuewe, T. Encrenaz, G. Wiedeman & M. Pizarro/ESO)

The effects of the impacts were seen over most of the electromagnetic spectrum. This is one of the shortest-wavelength observations, taken with the Extreme Ultraviolet Explorer (EUVE) satellite in Earth orbit at a wavelength of 58.4 nm where helium gas glows brightly. Normally, Jupiter is virtually invisible at this wavelength (left panel) because helium is hidden by the lighter hydrogen that floats on top of it in Jupiter's uppermost atmosphere. However, the impacts apparently stirred up enough helium into the outer atmosphere so that it was no longer masked by the hydrogen and produced a bright glow (right panel). Jupiter's disk is not resolved in these very low resolution images. (Randy Gladstone & the EUVE SL9 Observing Team/ UC Berkeley Center for Extreme Ultraviolet Astrophyics)

Impact Parameters for Fragments of Comet P/Shoemaker-Levy 9

Fragment	Date July	Accepted Time (UT)	Uncertainty (min)	Jovicentric Lat. (deg)	Long. (deg)	Merid. Angle (deg)	Angle E–J–F (deg)	Approx. Impact Brightness
A=21	16	20:11	4	-43.15	185	64.48	98.72	2
B=20	17	02:53	4	-43.17	69	63.82	99.18	0
C=19	17	07:12	4	-43.38	224	65.24	98.11	2
D=18	17	11:54	3	-43.46	34	65.58	97.85	2
E=17	17	15:11	3	-43.48	153	65.69	97.77	3
F=16	18	00:33	5	-43.55	134	64.46	98.62	0
G=15	18	07:33:32	1	-43.60	26	66.65	97.05	5
H=14	18	19:31:59	1	-43.74	100	66.85	96.88	4
K=12	19	10:24:14	1	-43.80	279	67.75	96.22	5
L=11	19	22:16:48	1	-43.92	49	68.19	95.87	5
N=9	20	10:29:17	1	-44.30	72	67.77	96.10	1
P2=8b	20	15:23	7	-44.64	250	67.08	96.52	0
Q2=7b	20	19:44	6	-44.26	46	69.45	94.91	1
Q1=7a	20	20:13	3	-44.05	63	69.55	94.87	3
R=6	21	05:34	3	-44.07	45	69.62	94.81	3
S=5	21	15:15	5	-44.16	33	69.95	94.56	4
T=4	21	18:10	7	-44.99	141	67.34	96.25	0
U=3	21	21:55	7	-44.43	276	68.82	95.32	1
V=2	22	04:23	5	-44.43	150	69.50	94.83	1
W=1	22	08:06:12	1	-44.15	283	70.63	94.08	3

Explanation of the table

The impact date and time is the time the impact would have been seen from Earth if the limb of Jupiter had not been in the way, that is the time of impact plus the light travel time to Earth.

The impact latitude is jovicentric, i.e. measured with reference to the center of Jupiter. The impact longitude is System III, measured westwards on the planet.

The meridian angle is the jovicentric longitude of impact measured from the midnight meridian towards the morning terminator. This relative longitude is known much more accurately than the absolute longitude.

Angle E–J–F is the Earth–Jupiter–Fragment angle at impact. Values greater than 90° indicate a farside impact.

Approximate impact brightness is a subjective rating of the brightness of each impact, from 0 (not detected) to 5 (brightest).

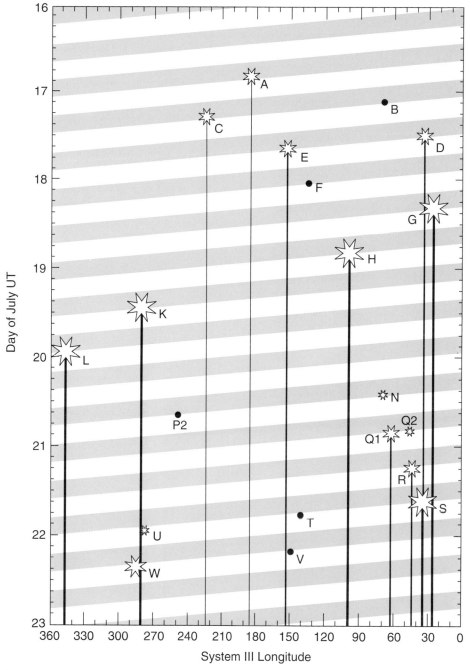

Impact week at a glance. The times and Jupiter longitudes of each impact are shown with star symbols whose sizes give a rough indication of the brightness of each flash. The black lines originating at most impacts show the longitude and approximate size of the persistent impact cloud left by the impact, thicker lines indicating the larger and more conspicuous clouds. The diagonal bands indicate the visibility of each part of Jupiter as the planet rotates: longitudes on the visible, daytime side of the planet are shown in white and the invisible night hemisphere is shown shaded. To determine Jupiter's appearance at any time during impact week, find that time on the vertical axis and place a horizontal ruler across the graph at that point. The visible side of Jupiter is marked by the longitudes where the ruler falls on a white background (possibly split between the left and right side of the graph), and the impact sites visible at that time, and their relative positions from left to right across the disk, will be shown by the points where the black vertical lines intersect the ruler.

A typical impact

During Impact Week there was little time to assimilate what was going on, but in the weeks and months that followed, the dominant theme in the e-mail traffic that continued to circulate around the globe changed from 'Look what we saw!' to 'What actually happened?'. No single observing team saw more than a small sample of the full range of phenomena. Comparison of observations taken in many different ways at many different wavelengths would be needed to build up the Big Picture. By the time of the annual meeting of the Division of Planetary Sciences of the American Astronomical Society, held in Bethesda, Maryland at the beginning of November 1994, and a subsequent workshop (hosted by the European Southern Observatory) in Germany in February 1995, this process of synthesis had come a long way. Astronomers had put together a pretty complete descriptive narrative of the sequence of events, though they were still groping towards a physical understanding of what they had seen.

In the images that follow we will illustrate that narrative by showing what happened during a typical impact, drawing on observations of the demise of many different fragments, seen in many different ways. To help clarify the sequence of events, we give the time of each image (where appropriate) relative to the impact from which it originated, taken to have been at T=0.

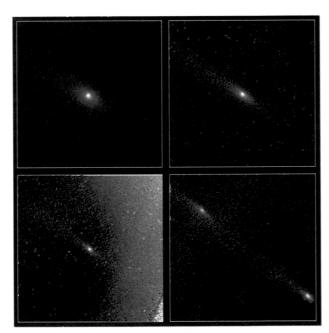

The comet in its final hours. Jovian tidal forces caused extreme stretching of the comae but the nuclei seemed to hold together: the bright cores of each fragment remained intact till at least ten hours before impact, when these last HST images were taken. Upper left: G, four days before impact. Upper right: K, 18 hours before impact. Lower left: L, 10 hours before impact. Lower right: Q1 and Q2, ten hours before impact. (Courtesy H. Weaver & T. E. Smith/NASA)

An artist's impression of the *Galileo* spacecraft entering Jupiter orbit in December 1995, at the start of a two-year investigation of the giant planet. By extreme good fortune, the spacecraft was suitably positioned to obtain direct views of the actual impacts while en route to its own jovian rendezvous. Its collection of instruments, designed for observations of the jovian system, also proved ideal for study of the impacts. *Galileo* was able to obtain images, light curves, and spectra of many of the impacts over a large range of wavelengths. The major difficulty with these observations was the fact that *Galileo*'s umbrella-like high-gain antenna, which was meant to be the main communication link to Earth, had failed to unfurl properly, so the observations of the comet impact had to be stored onboard and trickled back to Earth over several months through the much less powerful low-gain antenna. However, the observations were ingeniously designed to make the most of the limited downlink capability, and an impressive amount of science was packed into the small amount of data that could be returned. (Courtesy Jan Ludwinski, JPL/NASA)

(T=0) The *Galileo* spacecraft's view of the entry flash of the W impact, returned to Earth in mid-August. The images are part of a time sequence taken in green light at two-and-a-third second intervals. These cover seven seconds between 08:06:10 and 08:06:17 on 22 July (Earth equivalent). The flash brightened in a few seconds, saturating the camera (so the peak brightness is greater than it appears here), and then faded rapidly. This sudden flash was presumably produced by the first entry of the comet fragment into the atmosphere. Earth-based observers with small telescopes would have seen something like this if the impacts had occurred on the dayside: the brief flash would have been easily visible to the eye with very modest optical aid.

Strangely, several Earth-based telescopes saw faint 'precursor' infrared flashes that started several tens of seconds *before* the sudden *Galileo* flashes, despite *Galileo*'s more direct view. This is difficult to explain: perhaps the early flashes seen from Earth were produced by the comet fragments extremely high in the atmosphere, a long time before they exploded in the lower atmosphere, or perhaps they were created by small fragments or dust in advance of each main fragment. It is even possible that the sudden flashes seen by *Galileo* were not due to the comet entry at all, but rather the subsequent appearance of the fireball. With luck, continued comparison of the Galileo and Earth-based observations can resolve this puzzle. (JPL/NASA)

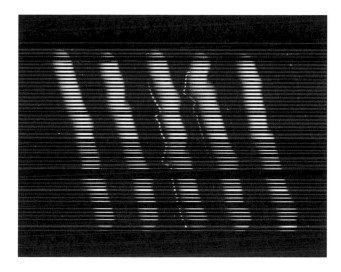

(T=0) In a successful effort to get better time resolution, several impacts were observed by *Galileo* in 'drift scan' mode. The camera was moved during each exposure so that Jupiter was smeared out across the CCD, and this process was repeated several times per image to pack as much information as possible onto a single frame. In this view of the K impact, the impact flash appears as a streak next to Jupiter in two of the middle scans, and the variation in brightness of the streak along its length can be translated into a variation in the flash brightness with time. Occasional wiggles in the scan are caused by jitter in the camera pointing. Fainter streaks are produced by the Galilean satellites. The black horizontal stripes represent parts of the image that were not sent back to Earth, in order to reduce the total amount of data that had to be returned through the very slow low-gain antenna.

This sequence (10:23:10 – 10:25:37 UT on 19 July) shows that the K impact produced a flash that began at 10:24:14 UT (Earth equivalent), brightened in a few seconds, as fast as for the impact of the fainter W fragment, but then lasted much longer – about 35 seconds. In general, the fragments that were brighter before impact tended to produce longer-lasting flashes. It seemed that for the larger impacts, the flash of the superheated air as the comet entered the atmosphere merged seamlessly with the beginning of the fireball as the hot gases began to rush back up the entry path. Peak brightness was about 10% of the brightness of Jupiter in this methane filter: too faint to produce the hoped-for visible illumination of the Galilean satellites. (M. Belton, T. Johnson, C. Chapman, K. Klaasen, C. Heffernan, J. Veverka, A. Ingersoll & *Galileo* Imaging Team)

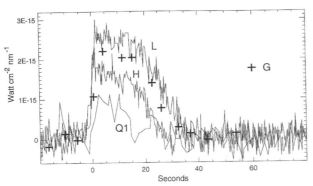

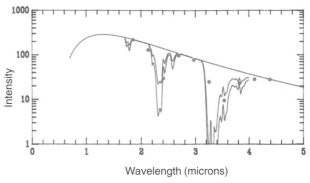

(T=0) The *Galileo* team chose to observe different impacts with different combinations of the spacecraft's suite of instruments. Impacts B, H, L and Q1 belonged to the photopolarimeter/radiometer (PPR) instrument, which did not take images of Jupiter but simply recorded its total brightness every 0.23 seconds at a wavelength of 945 nm. PPR also shared the G event with the NIMS instrument which scanned back and forth across Jupiter, allowing the PPR instrument to glimpse the impact every 5 seconds. Like Earth-based observers, PPR saw no sign of the B impact, but the others were clearly seen as a brightening of Jupiter amounting to a few percent. The G, H, L and Q1 events are shown here to the same scale. L was the brightest. The time history of the flash is the same as was seen by the *Galileo* camera: a rapid rise lasting a few seconds, a plateau, and a fading that lasted tens of seconds. The shapes of the brightest flashes were all remarkably similar. (G: 7:33:32 on 18 July; H: 19:31:59 on 18 July; L: 22:16:48 on 19 July; Q1: 20:13 on 20 July). (T. Martin, L. Tamppari & *Galileo* PPR team)

(T<1 minute) The most direct and detailed information on the initial growth of the impact fireballs that followed the entry flashes came from a third *Galileo* instrument, the Near-Infrared Mapping Spectrometer (NIMS). NIMS measured the brightness of the fireball at 17 wavelengths between 1 and 5 microns every 5 seconds during the G impact around 07:34 UT on 18 July, providing a crude spectrum that carried information on the changing size, temperature, and altitude of the fireball. The green dots show a spectrum obtained 21 seconds after the G fireball was first detected. This is matched by the radiation expected from an object 40 km across at a temperature of 2200 K or 3500° F (think for a moment about what this means!), shown by the three solid red curves. Because the fireball's radiation was absorbed by methane in the still-undisturbed atmosphere above it, we can estimate the altitude of the top of the fireball by measuring the quantity of overlying methane from the strength of the methane absorptions in the fireball spectrum. The three comparison spectra have been calculated assuming that the top of the fireball is above the atmosphere (no methane absorption bands), at 50 mbar pressure (moderate absorption bands), and at 100 mbar pressure (strong absorption bands). The shape of the actual spectrum suggests an altitude of perhaps 70 mbars. At this moment the fireball was radiating heat at a rate of several hundred million megawatts. (R. Carlson, P. Weissmann & NIMS team)

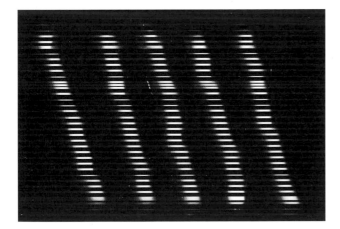

(T=0) The N impact flash in the 889-nm methane band, observed between 10:28:13 and 10:30:40 UT (Earth equivalent) on 20 July by the *Galileo* spacecraft using the 'drift scan' mode. As with the W impact, the N flash was brief, lasting for about 14 seconds. Peak brightness was about half that of the K impact. The impact flash appears as a short streak to the left of the central smeared Jupiter image. The flash began at 10:29:17 UT (Earth equivalent). (C. Chapman and B. Merline, & *Galileo* Imaging Team)

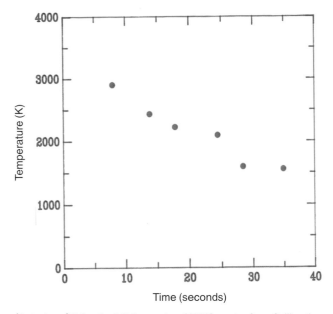

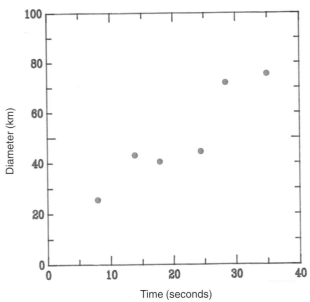

(T<1 minute) Using the full time series of NIMS spectra from *Galileo*, the cooling and expansion of the G fireball (around 07:34 UT on 18 July) can be followed. In the first 40 seconds it cooled from 3000 K to 1500 K (left-hand graph) and simultaneously expanded from 20 km to 80 km (graph on the right), a rate of 1 km per second, or about twice the speed of sound. Expansion was probably quite smooth. (The plots look irregular because of measurement errors.) NIMS spectra taken six minutes later showed the beginning of the infrared flash that was seen so brilliantly from Earth. This later flash was probably produced as the collapsing fireball slammed back into Jupiter's upper atmosphere. (R. Carlson, P. Weissmann & NIMS team)

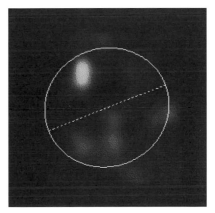

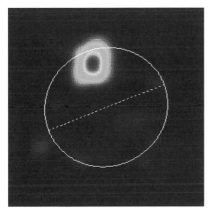

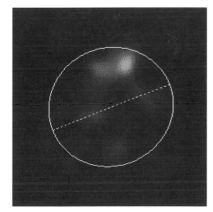

(T=−90 min, −3 min & +90 min) One of the most baffling observations made during impact week was this one at a wavelength of 1 nm from ROSAT, the German/US orbiting X-ray observatory. Between 08:34 and 09:17 UT on 19 July, 90 minutes before the K impact, Jupiter (shown by the white outline) was emitting few X-rays (left panel), but at 10:21 UT, three minutes *before* the K impact an intense burst of X-rays was seen coming from the northern hemisphere. Ninety minutes later (right panel, 11:45–12.30 UT), all is quiet again. It is very difficult to explain why X-rays would be produced before the impact rather than after, and some people have questioned the reality of the flash. Shortly before impact the K fragment passed close to the Io flux tube, the electrical current that flows between Io and Jupiter's poles, and this could conceivably have triggered the burst of X-rays, but this explanation seems far-fetched and there are few more palatable alternatives. (H. Waite, R. Gladstone & ROSAT SL9 observing team)

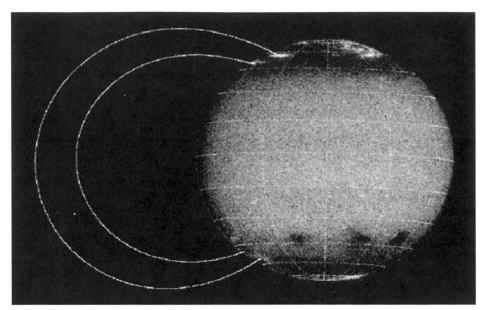

(T=45 min) One apparently strange aspect of the ROSAT X-ray data is that the flash is seen in the northern hemisphere while the impacts were in the southern hemisphere. However, a similar thing is seen more clearly in this remarkable HST image, taken at 10:09 UT on 19 July, 45 minutes *after* the K impact, at an ultraviolet wavelength (160 nm) in which Jupiter is faint but its aurorae glow brightly near the north and south poles. The C, A and E impact sites show as black spots due to the presence of absorbing dust high in the atmosphere. However the fresh K impact site (lower left) shows a faint glow, and if the lines of Jupiter's magnetic field (superimposed) are traced from the impact site around to the northern hemisphere a much brighter pair of spots can be seen, in a region where normally there are no aurorae at all.

Because this northern glow occurs after the impacts, an explanation is easier to come by than for the X-ray flash. Perhaps the impact released electrons into Jupiter's magnetosphere and these followed the field lines around to the northern hemisphere where they produced an auroral glow as they collided with the upper atmosphere. (John Clarke & HST UV imaging team)

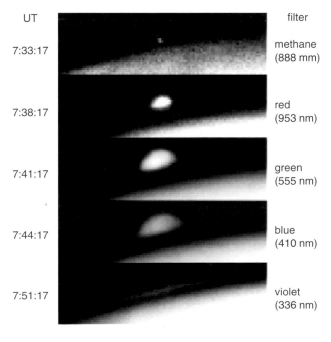

UT	filter
7:33:17	methane (888 mm)
7:38:17	red (953 nm)
7:41:17	green (555 nm)
7:44:17	blue (410 nm)
7:51:17	violet (336 nm)

(T=0–18 min) Among the most dramatic and informative HST images were the views of the expanding plumes as they rose and rotated from the hidden impact site into view and into sunlight. It was fortunate that the impacts occurred in the right place to afford this privileged sideways view. Even better, the position of the Earth relative to the Sun allowed us to look a little bit 'around the corner' onto the night side, so we could see the glow of the hot material in the fireball before it rose into sunlight. This sequence of the G impact plume is one of the best. *Galileo* observations tell us that the impact occurred at 7:33:32 UT. The first frame was exposed between 7:33:17 and 7:33:47, and thus includes the moment of impact. It already shows a bright spot glowing in Jupiter's shadow. As the impact site is out of view, it is not clear what we are seeing here: possibly the flash of the hidden impact reflected in cometary dust that has not yet hit, or a very high altitude bolide. By 7:38:17 UT, the top of plume has risen into sunlight, but glowing material can still be seen in Jupiter's shadow. By 7:41:17 UT the plume reached a peak altitude of about 3000 km, and began to collapse, though Jupiter's rotation continued to bring it into better view. By 7:51:17 collapse of the plume was almost complete and it appears as a thin 'pancake', though it is still high enough above the cloudtops that Jupiter's shadow can be seen cutting across its lower portion. (Space Telescope Science Institute)

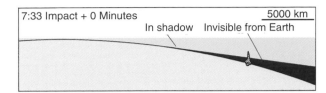

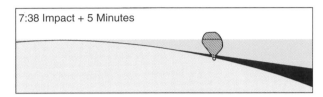

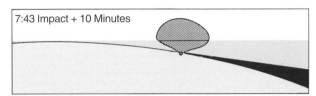

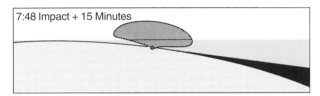

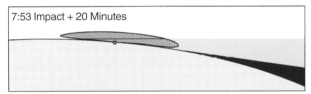

Enhancement of two of the HST images of the fireball developing after the G impact , shows material glowing in Jupiter's shadow below the bright sunlit part of the plume. The glow appears to be aligned along the direction of the comet's entry into the atmosphere, probably because the fireball escaped by this same route. (Space Telescope Science Institute)

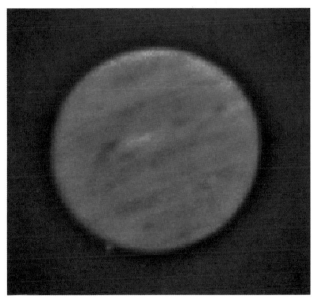

The explanation of HST's view of the G impact fireball, shown to scale. The view is a series of slices through Jupiter's limb, with the Sun to the left so that Jupiter casts a shadow into space towards the right. Earth is in a direction similar to the Sun's but is a little 'higher in the sky', so Earth sees the impact site before the Sun does and the region that is invisible from Earth is quite small. The approximate shape of the plume is shown as it grows and collapses. The glow seen from Earth at the moment of impact must be high above the cloudtops to be visible, but it is still well within Jupiter's shadow. The plume grows so fast that its upper part reaches sunlight less than five minutes after impact, though the impact site does not reach sunlight till about 25 minutes later. Universal times are shown for the G impact, to aid comparison with the HST images. The geometry is correct for the G impact but was similar for the others. (John Spencer/P. Doherty)

(T=10 min) The impact plumes were large enough to be visible from some ground-based telescopes. This view at a wavelength of 907 nm was taken with the 1-m Jacobus Kapteyn Telescope on La Palma 10 minutes after the L impact and clearly shows the impact plume off the limb of the planet: the plume appeared 6 minutes after impact and persisted for 13 minutes. Like the HST images, this view probably shows mostly sunlight reflected off the cloud of debris in the plume, rather than the brighter heat radiation from the plume that was seen at longer wavelengths. Spectra of the plume revealed the presence of sodium and other metals, presumably derived from the vaporised comet. (Alan Fitzsimmons)

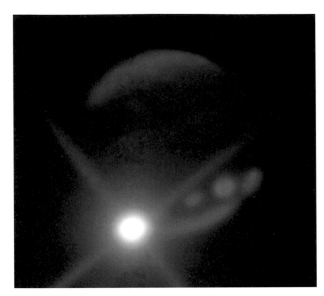

(T=10 min) The impacts were most spectacular in the infrared, where Earth-based telescopes saw the energy radiated by the clouds of hot gas and dust kicked up by the impacts, rather than just sunlight reflected off the dust. This infrared image shortly after the impact of fragment K was taken at 2.34 microns with the 2.3-m telescope of the Australian National University. The old scars of the A, E, and H impacts are also visible. (P. McGregor and M. Allen/ANU)

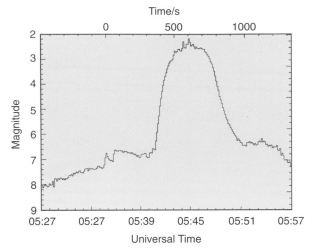

The light curve (brightness variation with time) of the R impact obtained from Keck telescope images at 2.3 microns. The brightness is measured using the astronomical magnitude scale, in which a magnitude *decrease* of 1.0 corresponds to a brightness *increase* by a factor of 2.5, and a magnitude decrease of 5 represents a hundred-fold brightness increase. Impact occurred at 05:35 UT. The scale across the top records seconds after impact. At the moment of impact there is a small, sharp, flash of light, followed a minute later by a second sudden pulse which persists for several minutes. Nearly six minutes after impact the overwhelmingly bright third pulse of light begins and brightens rapidly. Ten minutes after impact it is a hundred times brighter than the initial flash, then it fades away over the next few minutes. The first flash must have been produced by the entry of the comet into the atmosphere and it occurred 30 seconds before the *Galileo* spacecraft, with its direct view, saw anything at all. The second pulse of light is probably the first appearance of the hot fireball over Jupiter's horizon. The third and brightest occurs as the fireball collapses and slams back onto the atmosphere below. It is so bright because the collapse occurs over a very large area and also because the impact site has now rotated into more direct view. (Courtesy Imke de Pater)

(T= −7 min to +23 min) This exceptionally high-resolution view of the R impact was taken with the 10-m Keck telescope, Mauna Kea, at 2.3 microns, a wavelength at which most of Jupiter looks dark due to methane absorption. The first frame shows only the old G, L and K impact sites, appearing bright because they are above most of the absorbing methane in Jupiter's atmosphere. Seven seconds later, at 05:34:44, there is the first sign of a flash on Jupiter's limb, next to the old G site. A second flash is visible at 05:35:46. The much brighter main flash begins near 5:40:57 and is brilliant at 05:44:57. By 05:57:28 the flash has subsided but the new impact site can be seen rotating into sunlight, clearly separated from the nearby G site. (Imke de Pater)

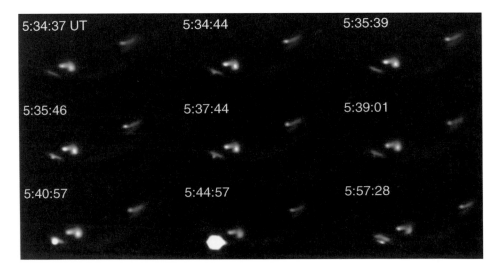

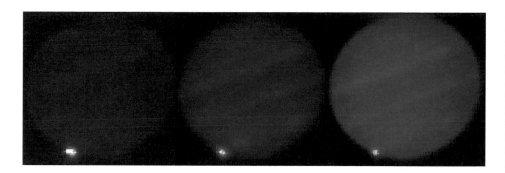

(T= 8 min) The impacts were also very bright at longer infrared wavelengths, where reflected sunlight is always negligible and both Jupiter and the impact sites are seen by the glow of their intrinsic heat radiation. These images of the R impact at 7.85 (left), 10.30 (center) and 12.2 microns (right) were taken at 05:42 UT on 21 July at the NASA Infrared Telescope Facility. This long-wavelength flash coincided almost perfectly in time with the third and most intense flash seen at 2.3 microns with the nearby Keck telescope, and is presumably a result of the same phenomenon, probably fallback of the impact ejecta onto Jupiter's atmosphere. (Glenn Orton and the NASA/IRTF SL9 Science Team)

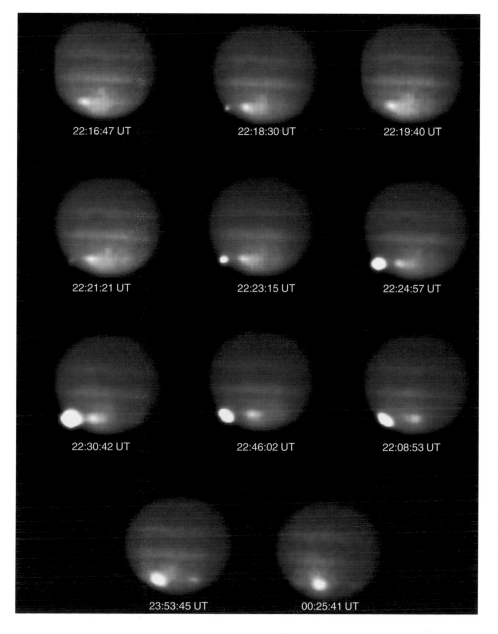

22:16:47 UT   22:18:30 UT   22:19:40 UT

22:21:21 UT   22:23:15 UT   22:24:57 UT

22:30:42 UT   22:46:02 UT   22:08:53 UT

23:53:45 UT   00:25:41 UT

(T = 0 min to 3.2 hours) A time series of the L impact seen at a wavelength of 12 microns with the CAMIRAS camera on the Nordic Optical Telescope on La Palma, Canary Islands. Impact was at 22:16:48 UT, as measured by *Galileo*. Two distinct flashes are seen. The first faint one, about three minutes after the impact, may correspond to the second flash seen at shorter infrared wavelengths, for instance at the R impact and may represent the first appearance of the fireball above the limb. The much brighter main flash lasts about 20 minutes but then lingers as a bright spot for several hours. In all these frames the K impact site is still glowing faintly 12 hours after impact. (Philippe Galdemard, CEA Service d'Astrophysique, Saclay, France, and the NOT impact observing team)

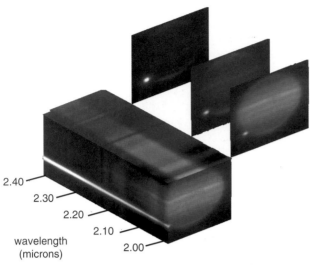

2.40
2.30
2.20
2.10
wavelength   2.00
(microns)

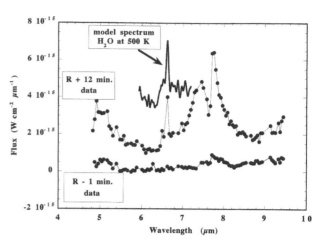

(T = 5 min) Most telescopes took either images or spectra of the impacts, but the 3.9-m Anglo-Australian Telescope got the best of both worlds by using an instrument called IRIS, a scanning spectrograph that produced 'image cubes' made of images at 128 different wavelengths, or if you prefer, spectra at 16 000 different points on Jupiter and the surrounding sky. This diagram attempts to illustrate the concept by showing a portion of one image cube taken at 07:38 UT on 18 July during the G impact. Three 'slices' have been taken out of the cube to show images of the impact at three different wavelengths, and along the sides of the cube can be seen the brightness of different regions of Jupiter and the sky at 128 wavelengths between 1.98 and 2.39 microns. The G impact plume appears as a very bright streak along the left side of the cube, and a plot of its brightness at each wavelength produces a spectrum. Cubes like this were taken every two minutes during many of the impacts. (David Crisp and Vikki Meadows/AAT; additional processing by John Spencer)

(T = 12 min) Astronomers were very anxious to look for the signature of water in the spectra of the impact sites. Water is frozen out in Jupiter's upper atmosphere and is hidden below the visible cloud layers. Water might have been added to the atmosphere from ice in the comet, or excavated from the lower atmosphere if the impacts went deep enough, or it might be cooked up during the impact from the combination of jovian hydrogen with cometary oxygen. Because jovian water is difficult to detect through the water in Earth's lower atmosphere, the first definitive observations of water were made from the Kuiper Airborne Observatory, flying at 41 000 feet. This spectrum of the collapsing R fireball, taken at 05:46 UT on 21 July, 12 minutes after impact, shows a strong peak at 6.6 microns due to water vapor at a temperature of 500 K, as well as emission from methane (CH_4) and sulfur dioxide (SO_2) between 7 and 8 microns. The top curve shows a theoretical spectrum of water at 500 K, which is an excellent match to the real spectrum. The bottom spectrum shows Jupiter's very faint appearance at these wavelengths before the impact. Using similar observations, Gordon Bjoraker estimated that the amount of water seen in the G impact corresponded to a 'one-kilometer ice cube', though this number is uncertain. SO_2 was not expected at all, having never been detected on Jupiter before. It was also seen by the HST. (Anne Sprague and the KAO HIFOGS team)

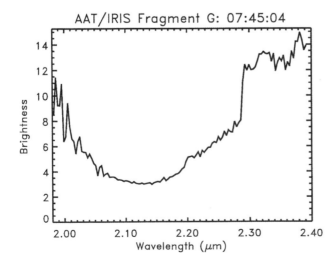

(T = 8 min) A spectrum of the probable collapse of the G impact fireball back onto the atmosphere, taken with the Anglo-Australian Telescope at 7:45 UT on 18 July near the peak of the impact flash. The spectrum was extracted from an IRIS 'image cube'. The sudden brightness increase at 2.30 microns is the signature of carbon monoxide (CO), probably observed as it was being created, perhaps by the reaction of carbon from Jupiter's atmosphere and oxygen from the comet. Emission peaks near 2.0 and 2.1 microns are from water (H_2O) and hot ammonia (NH_3), respectively. Ammonia is not normally seen in the upper atmosphere and may have been excavated from lower levels by the impact. (David Crisp and Vikki Meadows/AAT)

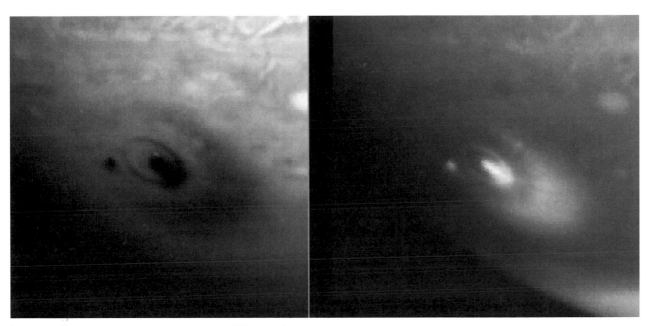

(T = 108 min) This pair of HST images, taken at 09:22 UT on 18 July, compares the appearance of the fresh G impact site in green light (550 nm, left) and through a filter sensitive to methane (890 nm, right). The intrinsically dark impact features appear bright at the methane wavelength because they are high in the atmosphere, above most of the methane absorption. The inner, complete, ring is an expanding wave of some kind, perhaps analogous to ripples on a pond, though we do not understand why it is visible. The central streak, and the outer crescent-shaped feature, appear to have been left by the obliquely-ejected fireball as it collapsed back onto Jupiter's atmosphere, though we do not understand the gap between the streak and the crescent. Another puzzling feature is the orientation of the whole pattern, which, though beautifully symmetrical, is not lined up perfectly with the direction of the incoming comet. The most likely explanation for the discrepancy is that the collapsing ejecta skidded across the top of the atmosphere for some time after re-impact before coming to rest, while the planet continued to rotate below it, so that the ejecta pattern was twisted relative to the planet. (H. Hammel, MIT/NASA)

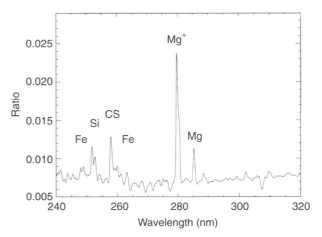

(T = 45 min) This ultraviolet HST spectrum of the combined G and S impact sites, taken at 16:00 UT on 21 July, 45 minutes after the S impact, shows emission from silicon, iron and magnesium – elements that have no place in a planetary atmosphere and could only have come from the comet. The intensity of the spectrum is shown relative to a pre-impact spectrum of Jupiter so only new features produced by the comet impact are seen. The iron and magnesium atoms are ionized (i.e. they have missing electrons). It is unlikely that these would have survived the three and a half days since the G impact so they probably came from the S fragment. (Keith Noll and the HST comet impact spectroscopy team)

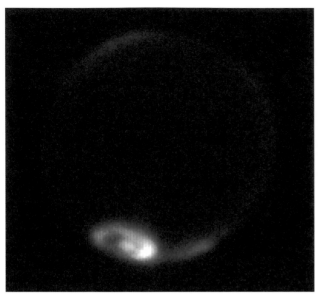

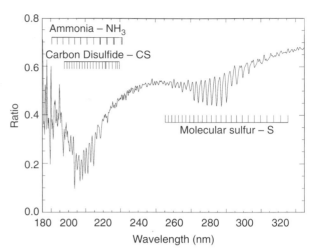

(T=3.5 hours)This ultraviolet spectrum of the G impact site, taken by the HST at about 11.00 UT on 18 July, was such a surprise that it took some time before the science team understood what they were seeing. Eventually they realised that sulfur compounds were responsible for most of the fine details in the spectrum. Sulfur atoms in pairs (S_2) produced the absorption centered at 270 nm, and carbon disulphide (CS_2) and ammonia (NH_3) produced the more complex absorptions near 200 nm. At first it was thought that this spectrum represented an amount of sulfur in the atmosphere exceeding the estimated total mass of the G fragment, implying that most of the sulfur must have come from Jupiter itself. However, the estimated abundance was revised downwards after more detailed calculations and it is possible that all the sulfur may have come from the comet after all. There is no sign of the emission from iron, magnesium or silicon that was seen 45 minutes after the S impact. It has probably faded away in the 3.5 hours since the G impact occurred. (Keith Noll and the HST SL9 spectroscopy team)

(T = 77 min) One of the most remarkable effects seen after the largest impacts was an enormous bright ring spreading out from the impact site. This was only visible at wavelengths between 3 and 4 microns, and few observatories had cameras that worked in this wavelength range. The observers at the Australian National University's 2.3-m telescope may have been the only people to look at the right wavelength at the right time to see this phenomenon. The ring seen here (imaged at 08:50 on 18 July at 3.28 microns) is approximately 33 000 km in diameter, much bigger than the 7500 km diameter wave seen 30 minutes later by the HST and even bigger than the asymmetric dark halo also seen in those HST images, which was about 12 000 to 20 000 km in diameter. It may have been far-flung ejecta re-impacting the atmosphere, or possibly some form of atmospheric wave. Visibility of the ring is enhanced by the extreme darkness of the rest of Jupiter at this wavelength, which is strongly absorbed by methane. The aurorae can be seen glowing at the north and south poles. (Peter McGregor and Mark Allen/ANU)

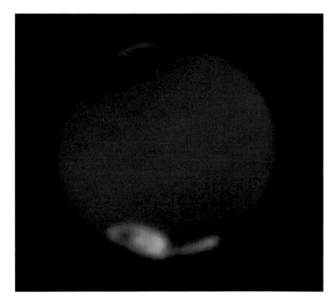

(T = 77 min) A 3-color image of the bright ring around the G impact site visible in the 3–4 micron wavelength band. It was taken at about 08:50 UT on 18 July with the 2.3-m telescope of the Australian National University. Methane absorption makes Jupiter very dark at 3.42 microns, which is represented by green in this color composite, so Jupiter's disk appears purple. The polar aurorae are bright at 3.42 microns and thus appear green. The impact ring is brightest at 3.09 microns (blue) and 3.42 microns, and thus appears in shades of blue and green. Red represents 3.99 microns. (Peter McGregor and Mark Allen/ANU)

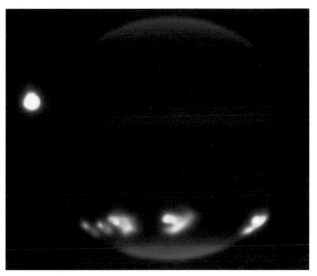

The visible appearance of two large impact scars. This image, taken on 20 July at 01:53 UT, shows the old G site (left) and the fresher L site (right), which is about three and a half hours old. This is a color composite of three exposures through different filters, taken with a 40-cm (16-inch) telescope. (Don Parker)

The intrinsically dark impact sites appeared bright at wavelengths where methane absorption in the atmosphere below the impact sites darkened the rest of the disk. This was a strong indication that the impact clouds were very high in the atmosphere, above most of the methane, at around the 1 millibar pressure level which is 150 km above the level of the 'normal' clouds. This is one of the highest resolution ground-based infrared (2.3-micron) images of the impact sites, taken at 06:50 UT on 21 July with the adaptive optics system on the University of Hawaii 2.2-m telescope on Mauna Kea. From left to right, names and ages of the visible impact sites are Q1 (10.6 hours), R (1.2 hours), G (71.3 hours), L (32.6 hours), and K (44.4 hours). (K. Hodapp, J. Hora, K. Jim & D. Jewitt; processed by R. Wainscoat and L. Cowie)

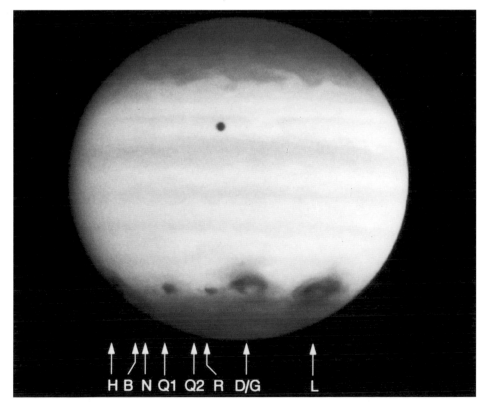

An ultraviolet (255 nm) image of Jupiter taken by the Hubble Space Telescope at 07:55 UT on 21 July, about two and a half hours after the R impact. Eight impact sites can be seen They are very dark in the ultraviolet because of the large quantity of dust deposited in the jovian stratosphere. The dark spot above the center of the planet is Io. (J. Clarke, University of Michigan/NASA)

The pair of full-disk images was obtained with the 0.8-m telescope of the University of Texas McDonald Observatory on 20 July at 03:02 and 03:06 UT. An 893 nm filter, which isolates absorption by methane gas, was used for the left image. This filter makes high cloud features appear bright. An 829 nm filter was used for the right hand image. The human eye would see a view similar to this. In the left image, the L impact site is on the right limb, the G and D impact sites (merged together) are just to the left of the L site, and the H site is near the east (left) limb. All appear as bright spots. The Great Red Spot can be seen just coming over the east limb. The right image shows only the G and H impact sites. The L

site is lost on the limb. (Chan Na, Southwest Research Institute, Wayne Pryor, University of Colorado, & Anita Cochran, University of Texas/McDonald Observatory)

The third image, showing the lower third of Jupiter only, is at 10.3 microns and was taken with the 5-m (200-inch) Hale Telescope at Palomar Observatory at 02:50 UT on 20 July, just a few minutes before the McDonald images. The same three impact sites are visible. (P. Nicholson, T. Hayward, J. Miles, C. McGhee, J. VanCleve (Cornell University); G. Neugebauer, K. Matthews, D. Shupe, A. Weinberger (Caltech))

The healing process

When the final fragment collided with Jupiter, the first impact site (A) was nearly six days old and still conspicuous. Everyone wondered how long the impact scars would last. We knew that the black clouds, whatever they were made of, were high in the stratosphere where there is no vertical movement to stir them down into the lower atmosphere. They would have to settle out slowly by gravity. At the same time, stratospheric winds would stir the cloud material horizontally and would be expected to break up and dilute the discrete clouds at each impact site. We would just have to wait and see how long these processes took before the face of Jupiter returned to normal. By monitoring the fading scars, astronomers hoped to gain new knowledge about how the jovian stratosphere works.

The HST obtained the most detailed views of the breakup of the impact sites. This image, taken between 06:22 and 06:28 UT on 21 July, shows the distortion of the G and L sites due to jovian stratospheric winds. The fresh R site, 50 minutes old, is just visible on the terminator and part of the K site can be seen near the opposite limb. (H. Hammel and the HST comet impact imaging team; additional processing by R. Evans, JPL; courtesy J. Trauger)

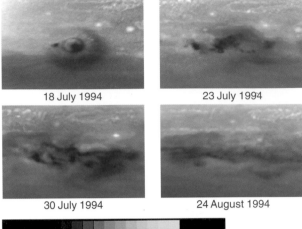

18 July 1994   23 July 1994

30 July 1994   24 August 1994

This HST sequence shows the disintegration of the G site, along with the smaller nearby D, R and S sites, over the course of a month. Each frame covers the same region of 40 degrees in latitude and 60 degrees in longitude. The R and S impacts occurred after the first frame but are visible in the second. S landed on top of D, and R is the detached spot to the left of the main complex. The impact clouds do not spread uniformly, but are torn into tendrils and knots. By late August they have faded significantly but are still conspicuous. (H. Hammel, MIT/NASA)

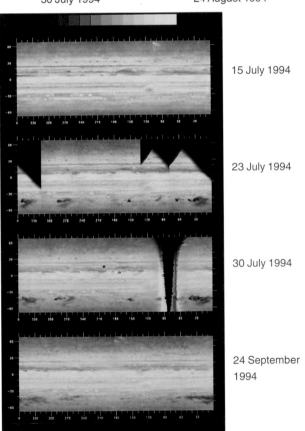

15 July 1994

23 July 1994

30 July 1994

24 September 1994

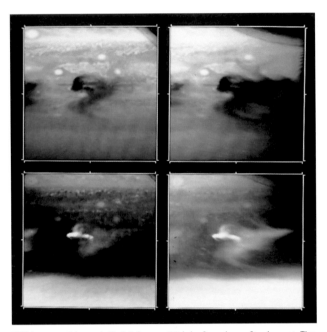

HST images of impact site L taken on 23 July, four days after impact. The impact cloud is being stretched into a Z-shaped pattern by the alternate east–west winds of the jovian stratosphere. This was the first time that the wind patterns in this normally clear part of the atmosphere could be seen. The upper pair of frames are taken in blue light, at a wavelength of 410 nm, and the sites appear dark. The lower pair are through the 888 nm methane filter, and the sites appear bright. The right-hand images were taken near the limb of Jupiter where the longer slanting path of the light through the impact clouds enhances their contrast, but have been corrected to show a vertical view. The cloud structure shows up particularly well in the methane-band image taken near the limb. (H. Hammel, MIT/NASA)

Four maps of Jupiter constructed from HST images, showing Jupiter's appearance, approximately in true color, before and immediately after the impacts, and over subsequent weeks. Gaps are due to incomplete coverage. In the second panel the most prominent impact sites, from left to right, are L, K, E, H, Q1 and the R/S/D/G complex. The blending and fading of the impact sites shows up clearly. By late August the R, S, D, G and L sites had merged into an irregular band covering 100 degrees of longitude, while the K/W, E, H and Q sites still maintained some individuality. The A and C sites had disappeared. At the level in the atmosphere of the clouds seen at visible wavelengths there was little spreading north or south, however. The southerly halos of the K and L sites were both stretched into elegant Z-shapes by the alternate east–west stratospheric winds. (H. Hammel, A. Ingersoll/NASA)

violet (336 nm)   far UV (160 nm)

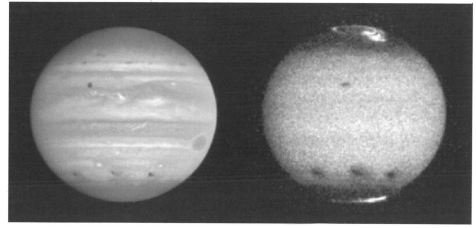

17 July 1994

29–30 July 1994

Thirty-nine days after the end of impact week there was still plenty of structure in the impact clouds, as seen in this high-resolution Keck Telescope image, taken at 06:20 UT on 30 August in the infrared (2.12 microns). The view is dominated by the scattered remains of the K impact site to the left of center, with the E site on the right limb and the L site, merging with K, appearing on the terminator on the left. There is hardly any sign of the C and A sites, which should be just right of center. Very little spread of material northward from the impact latitude can be seen. (Imke de Pater)

HST followed the evolution of the impact sites in the ultraviolet as well as the visible. These two pairs of images show the same face of Jupiter at two wavelengths, revealing the evolution of the relatively small C, A and E sites over a 13-day period from 17 July, a few hours after the E impact, to 30 July. The left-hand images, in the nearest ultraviolet, show Jupiter appearing much as it does in the visible. The sites of the three impacts appear as small dark spots on 17 July. The A and C sites have largely disappeared 13 days later, though the E site remains conspicuous. The right hand images, in the far ultraviolet, tell a different story. Light at this wavelength does not penetrate very deeply into Jupiter's atmosphere, so only the highest-altitude impact clouds can be seen. However, the clouds absorb very strongly in the ultraviolet so they can be seen in places where they are too thin to show up at longer wavelengths. On 17 July, for example, the impact sites appear larger in the far ultraviolet image. By 30 July, the far UV images shows that the impact sites have spread thousands of kilometers in an east–west direction and are starting to merge. There is also striking evidence of northward spreading. A dark cloud has appeared north of the A and E sites where none can be seen at longer wavelengths. Perhaps the northward winds exist only at the highest elevations in the atmosphere and thus only affect the very highest impact material, which is only seen in the far ultraviolet. The images also provide a magnificent view of the jovian aurorae at the north and south poles, and Io can be seen passing in front of Jupiter in the two near UV images. (J. Clarke, G. Ballester, J. Trauger/NASA)

E   H   Q   R G   L   K   C

24–25 July

17–19 August

28–29 August

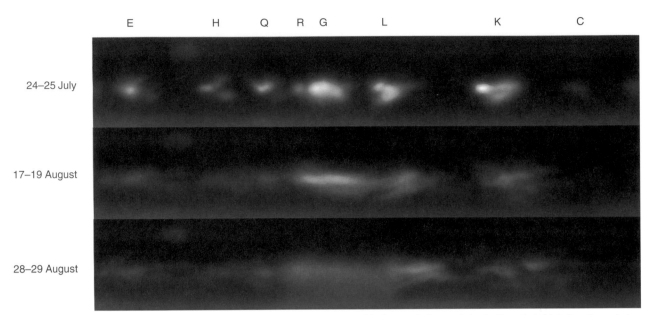

Jovian latitudes 20–70°S, from late July to late August, in the 2.3-micron wavelength methane band. The strips were constructed by piecing together four or five images from the 5-m (200-inch) Hale Telescope over a full rotation of Jupiter, on three different sets of dates. The spreading, merging and fading of the impact sites is well seen. By the end of August there was virtually a continuous band of material stretching around Jupiter at the impact latitude. Note that this infrared mosaic is centered differently from its visible counterpart. The A site is divided between the two ends of the mosaic. The Great Red Spot is dimly visible at the upper left of each strip. (P. Nicholson & J. Moersch, Cornell University; G. Neugebauer & K. Matthews, Caltech)

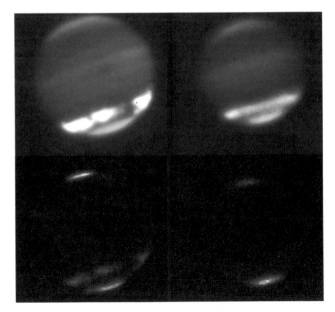

The dust settles. The impact clouds were sinking as well as spreading horizontally. All these images were taken with the 3.2-m NASA Infrared Telescope Facility, Mauna Kea, in methane bands, but the methane absorption is much stronger at 3.41 microns (the wavelength of the lower pair) than at 2.27 microns (upper pair), so the lower pair of images show the impact clouds at much higher altitudes. The impact sites, from left to right, are D/G/S, L and K/W. In the left hand images, taken on 7 August, the impact sites are clearly seen at 3.41 microns, indicating the presence of reflecting dust very high in the atmosphere, while two months later, on 11 October, the impact sites are still conspicuous at 2.27 microns (though fainter and very smeared out), but are essentially invisible at 3.41 microns. The atmosphere at the high altitudes seen at 3.41 microns is now clear of dust. Note the apparent shrinking of Jupiter between August and October: the Earth's distance from Jupiter has increased by 15% in this interval. Note also the polar aurorae at 3.41 microns. The northern aurora, which may have brightened as a result of the impacts, had faded to more usual levels by October. (K. Baines, J. Granahan & the IRTF comet collision science team)

After early October 1994, it was impossible to observe Jupiter because it passed behind the Sun, as viewed from Earth. With a few heroic exceptions, there was a two-month hiatus in observations. By mid-December Jupiter emerged into the pre-dawn sky. The first infrared images (2.3 microns) were claimed by the indefatigable Calar Alto team on the morning of December 19, and they obtained this view five days later at 07:00 UT on 24 December 1994. The R site is near the central meridian and the K site near the right limb, but the variations around the impact band are by now quite subtle and the sites have faded since October. Io is conspicuous to the right of Jupiter. (T. Herbst, P. Bizenberger, K. Reinsch, and V. Burwitz)

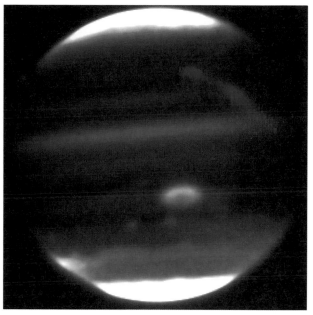

Nearly 8 months after the impacts, on 14 March 1995, there was still enough high-altitude dust in the jovian stratosphere to brighten the impact latitudes considerably compared to similar latitudes in the northern hemisphere as seen in this methane-band (2.3 micron) image from the NASA Infrared Telescope Facility, but the impact sites were then much fainter than the polar caps. Plenty of structure could still be seen – the wavy northern margin of the main impact cloud band, for instance – but by then this probably reflected stratospheric wind patterns and had little to do with the original distribution of dust. However, the brightening on the left limb might be associated with the K impact site. (John Spencer)

The amateur perspective, by John H. Rogers

Before the impacts, no-one knew what, if anything, would be visible through modest Earth-based telescopes. There were predictions that bright clouds, or expanding rings, or new oval spots might be visible to amateurs. But equally, small telescope owners were cautioned that they may well see nothing at all. Hardly anyone anticipated the huge black clouds that would become more conspicuous than anything else on the planet.

Nevertheless, amateurs worldwide were scrutinizing Jupiter during the impact week, looking for the fireballs themselves, their reflections from the jovian moons, and for any lasting effects the impacts might produce. In fact, there were only a few reports of fireball flashes being seen visually. None have subsequently been confirmed independently, though some sightings may have been real, as illustrated by the Hubble Space Telescope images in visible light of the fireball created by impact G. But the

dark spots were obvious. The largest were clearly visible even in telescopes with apertures as small as 6 cm, and remained so for well over a month.

As Director of the Jupiter Section of the British Astronomical Association (BAA), I received observations from observers in Belgium, Bolivia, Germany, India, Italy, Japan, Spain, Sweden and the USA, as well as from within the UK. The Association of Lunar and Planetary Observers (ALPO) collected reports in the USA, and the two organizations shared data. Some amateurs made intense special efforts to use large telescopes. In the UK, Patrick Moore got the moth-balled 26-inch refractor at Herstmonceux, former site of the Royal Greenwich Observatory, working for the occasion. Some obtained methane-band filters to use with their CCD cameras. Don Parker in Florida and Isao Miyazaki in Okinawa both used them in addition to making their usual high

17 July 1994
20:32 UT

18 July 1994
19:50 UT

19 July 1994
20:45 UT

20 July 1994
20:44 UT

21 July 1994
18:49 UT

22 July 1994
19:55 UT

23 July 1994
19:30 UT

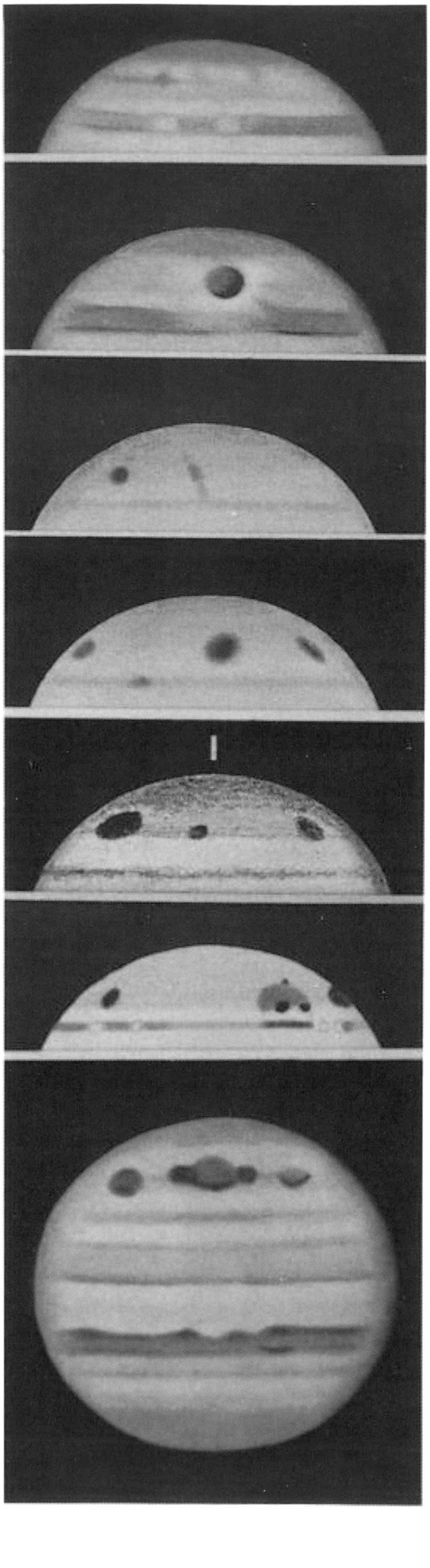

This Impact Week sequence consists of drawings made by amateur observers Lee Macdonald, David Gray, James Lancashire (UK) and Johan Warell (Uppsala, Sweden). They used instruments ranging from 22 cm to 42 cm in aperture. In keeping with the practice of amateur observers, these drawings (and the other illustrations in this section) have south at the top – the way the planet appears to a visual observer using an astronomical telescope, which inverts the image.

17 July, 20:32 UT (Gray). The C site is just visible.
18 July, 19:50 UT (Gray). Shows the G site.
19 July, 20:45 UT (Macdonald). The E site is dark and circular. The A site, also visible, is fading.
20 July, 20:44 UT (Macdonald). The sites visible are K, L and G.
21 July, 18:49 UT (Warell). Shows sites G, Q1 and H.
22 July, 19:55 UT (Lancashire). The large site is the K/W complex. C and L are also visible.
23 July, 19.30 UT (Gray). The G/S/R complex on the meridian, with the L and Q1 sites either side.

17 July 1994
17:00 UT

18 July 1994
14:45 UT

19 July 1994
13:30 UT

21 July 1994
13:35 UT

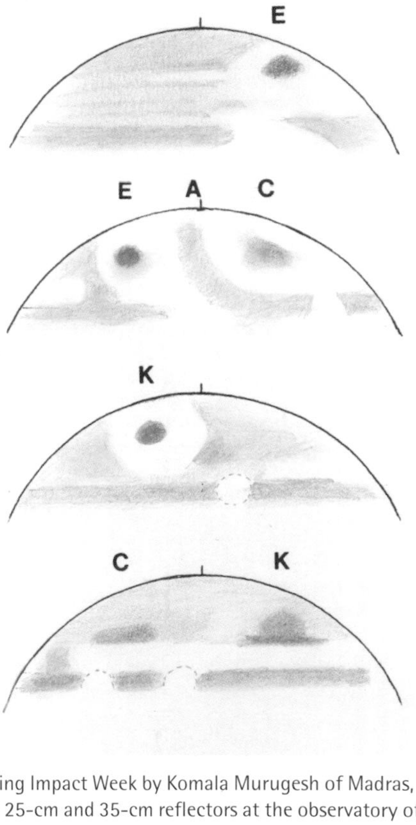

Drawings made during Impact Week by Komala Murugesh of Madras, India, who used the 25-cm and 35-cm reflectors at the observatory of P. Devada. 17 July, 17:00 UT: Site E on its first appearance. 18 July, 14:45 UT: Sites E, A and C. (The ring around the C site is an illusion created by the A site and the adjacent dark belts.) 19 July, 13.30 UT: Site K on its first passage. 21 July, 13:35 UT: Site C, now appearing as a streak, and site K. (South is at the top.)

resolution images. Others travelled south to see the planet at a higher altitude above the horizon, like Mark Bosselaers who went from Belgium to Tenerife. Meanwhile, thousands of other amateur astronomers and their guests looked at the impact sites with simple 'back-yard' telescopes.

The first impact occurred around sunset for western Europe on 16 July. With generally poor weather that evening, there were only a few suspected sightings of a dark spot as the impact site rotated onto the Earth-facing disk of Jupiter. However, as the sky darkened in Florida, this small dark spot where fragment A had struck was clearly seen by several amateurs. By the time the planet could be imaged from Okinawa, spot A on its second passage was followed by a second dark spot from the C impact. After another 10-hour rotation period had elapsed, both spots were evident to several European observers on the evening of 17 July, having clearly intensified during the first day.

Impacts D and E also produced dark scars. By contrast, the impacts of B and F, fragments that were displaced tailward from the main line of nuclei in the comet, produced little or no fireball and no visible scar.

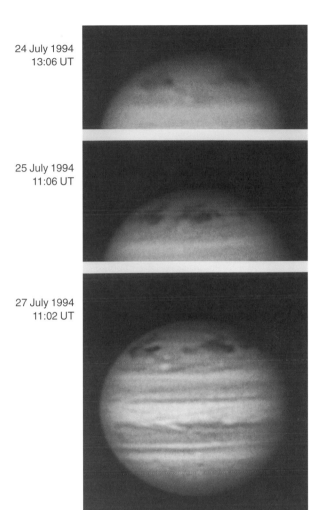

24 July 1994
13:06 UT

25 July 1994
11:06 UT

27 July 1994
11:02 UT

These images of Jupiter in the week after the impacts were made by amateur observer, Isao Miyazaki in Okinawa, Japan. He used a CCD camera on a 40-cm reflector. 24 July, 13:06 UT: The K/W complex (on the eastern side) and L sites. 25 July, 11:06 UT: The D/G/S/R complex is across the central meridian. Site L is on the eastern side and site Q1 towards the west. 27 July, 11:02 UT: Site K/W is on the eastern side. The dark core is clearly double. To the north of this site are three pre-existing white ovals. The other site in view, L, has emitted a dusky patch on its eastern side and long streamers to the south; it spans latitudes 43–66°S. (South is at the top.)

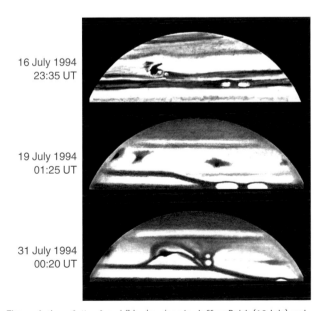

16 July 1994
23:35 UT

19 July 1994
01:25 UT

31 July 1994
00:20 UT

The evolution of sites A and C in drawings by Jeffrey Beish (16 July) and Carlos Hernandez (19 and 31 July). 16 July, 23:35 UT: The first distinct visual sighting of impact site A on its first passage. 19 July, 01:25 UT: Site A is fading, site C is near the central meridian and site E on the eastern limb. 31 July, 00.20 UT: Site C is the faint streak near the western limb, site A has virtually disappeared and site E is the conspicuous dark spot. (South is at the top.)

A pattern emerged that off-line nuclei 'fizzled', whereas in-line nuclei produced dark clouds that were visible as soon as the site came round the limb after the impact. These clouds tended to appear larger and/or darker the next day. The size of each visible scar was roughly in proportion to the reported magnitude of the fireball. When near the limb, these clouds were at least as dark as when fully on the disk, an effect of their very high altitude.

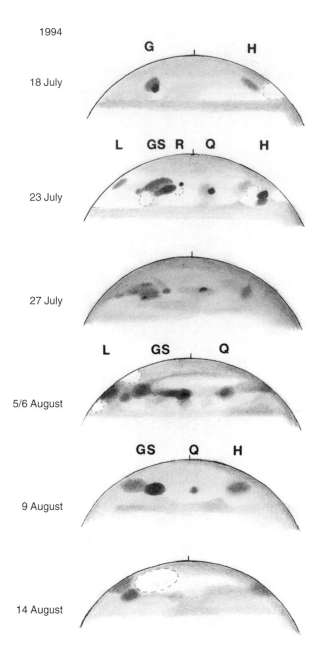

1994

18 July

23 July

27 July

5/6 August

9 August

14 August

The evolution of the impact scars in the region between sites G and H recorded in a series of drawings by John Rogers. He used a 30-cm refractor in Cambridge, UK, except for the 27 July drawing (copied from a CCD image by Isao Miyazaki, Japan) and the 5/6 August drawing (copied from a CCD image by Joel Stuckey, New York, USA). (South is at the top.)

It was on 18 July, with the impacts of the larger nuclei G and H, that the full spectacle started to unfold. That evening, site G was a stunning sight for observers in Europe. Site H was already quite dark as it rotated on, an hour after the impact. Fragments K and L produced equally great dark scars on 19 July.

So there was great anticipation on the evening of 20 July, when the double fragment Q, formerly the brightest piece of all, was due to hit. The great sites K, L and G were parading across the disk, each now about as big as the Great Red Spot, equivalent to several times the diameter of the Earth. They were intensely dark, with a black spot in one corner. In the event, not only did Q2 and Q1 fail to produce visible flashes, but the only new scar was rather a faint one for Q1. Q1 turned out to be the first example of a third type of fragment, which produced briefer fireballs and smaller dark spots than earlier fragments of comparable brightness. The next day, impact R was similar. S, U and W, piling into the pre-existing scars of G and K, did not obviously add to them, although these great spots were becoming larger and more complex.

By 22–25 July, when the last impacts had taken place, the planet looked quite unlike it had ever done before. While sites A and C were fading on one side, on the other side, sites K/W, L and D/G/S were immense, with sites R, Q1, H and E following. They were developing in shape and intensity as local winds spread the black 'smoke' this way and that. We could see half a dozen impact scars spattered across the disk – black and grey, large and small, northerly and southerly – and the line of impacts was even more conspicuous than the major cloud belts. The planet certainly looked as if it had been thoroughly carpet-bombed.

Over the next month, the keenest amateurs continued to track the impact sites. Most observers do not have imaging equipment and draw the visible features as they view. The results depend to some extent on the artistic skill and accuracy of the individual. Amateurs can also measure the longitudes of features, by timing their transit across the center-line (the central meridian) of the planet's disk. Observers experienced increasing difficulty as the planet's path through the sky took it closer to the Sun and it began to sink into evening twilight, but they continued until mid-September from Florida and up to mid-October from India.

Visible evidence of most sites lasted over a month, although site A disappeared within two weeks. Typically, they were beginning to fade after 10–20 days, while grey patches were appearing outside the previous boundaries of some sites. Images by Isao Miyazaki and Don Parker showed long dark streamers stretching out from site L. By late August, sites A and C were gone from view and sites Q1, H and E were reduced to indistinct condensations on a dusky belt, which extended all round the planet at the

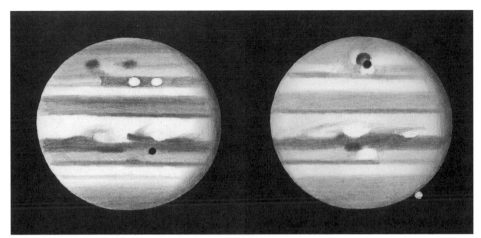

Drawings of the scarred face of Jupiter in July and August 1994 made by John Rogers, using a 30-cm refractor in Cambridge, UK. 17 July, 19:52 UT; 18 July, 20:07 UT; 29 July, 20:4 UT; 8 August, 19:43 UT. (South is at the top.)

17 July, 19:52 UT 18 July, 20:07 UT

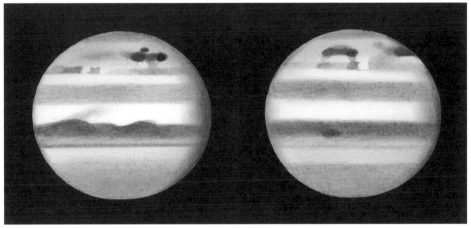

29 July, 20:34 UT 8 August, 19:43 UT

impact latitude. The three largest complexes – K/W, L and D/G/S – were still very large and dark in late August but had changed shape. The impression as late as mid-August that some dark sites still appeared darkest when near the limb suggested that lot of the material remained in the stratosphere even then. This is confirmed by amateur images in the 0.89-micron methane band, which still showed the major impact sites and the developing belt as bright features.

At first there were no systematic motions relative to the local current for the region (the SSS Temperate Current), which happens to have the same rotation period as the planet's deep interior (System III). There was evidence for a variety of local motions, such as an increase followed by a decrease in the longitude of site E, eastward and westward expansion of site K/W, and northward extension of site H. As the black cores

dissipated, several of the sites began to spread towards lower longitudes and, by late August, dark material had spread for tens of degrees in front of the original boundaries of the K/W and L impact sites.

Amateur astronomers not only had a week of fantastic excitement but were able to produce a remarkably detailed record of the individual impacts and the evolution of the sites afterwards. Limited only by the weather, not the allocation of telescope time, amateurs could follow each night as the whole sequence of events unfolded. And apart from the excitement and the science, we saw something to change our view of the solar system. Planetary catastrophes responsible for gouging out craters, shattering moons or decimating life millions of years ago, were previously only inferred or imagined. Now we have seen such a catastrophe happening in front of our very eyes.

9

Wiser after the Event?

H. Jay Melosh

Months after the crash, astronomers were still puzzling over some of the major questions about the comet and its impact. How big were the fragments and their parent comet? How deep did the pieces penetrate into Jupiter's atmosphere? What were the dark clouds made of? Answers were beginning to emerge, but interpretation of the observations proved less than straight-forward.

Comet Crash Week, 16–22 July 1994, will remain forever in the memories of astronomers and lay people alike as a period of excitement and delight. In spite of an abundance of observations from a vast array of instruments, and the unprecedented speed at which astronomers shared their observations through electronic networks, no one really understood what was being seen during that glorious and hectic week.

As I write, seven months after the impacts, observers and theorists have had a chance to consider the data and construct models to explain it. Based on past experience with space missions, most of us believed that, by the time six months had passed, the main outlines of the events of Comet Crash Week would have been drawn and a consensus on the principal features of the impacts themselves would have been reached. However, Comet Shoemaker–Levy 9, although no longer physically present in the universe, is still causing consternation and confusion on Earth.

At the meeting of the American Geophysical Union held in San Francisco during the first week of December 1994, a special session on the impact of SL9 showed that a consensus was still far away on some points, although agreement seemed to have been reached on others. The most uncertain aspect was still the size of the comet fragments that struck the atmosphere of Jupiter. Estimates of the maximum fragment size ranged from a diameter of 100 m to about 4 km – if anything, an even wider range than was proposed before the impacts actually occurred! The depth to which the fragments had penetrated was likewise uncertain, although a majority of observers believed that the penetration was much shallower than predicted beforehand. On the other hand, some agreement had been reached on the sequence of plume formation and of the timing of the different phases of plume growth. It is clear that almost everyone was surprised at the sheer size of the plumes and at the persistence of the dark spots that developed around the impact sites.

How big was Comet Shoemaker-Levy 9?

The debate about the size of SL9 began almost as soon as it was found. After its serendipitous discovery, followed by Jim Scotti's famous image made with the Spacewatch telescope, the Hubble Space Telescope was brought into play to produce high resolution images. Unfortunately, even the superior resolution of the HST could not reveal the actual comet nuclei. The smallest object it can image in the vicinity of Jupiter is larger than 100 km. Since even the maximum estimates for the size of fragments were a diameter of 10 km, it was clear from the beginning that more indirect means had to be used to determine their real dimensions.

An upper limit to the parent comet's size was provided by a negative observation. In spite of many efforts to find pre-discovery images of the comet before it made its first close pass by Jupiter in July 1992, no such image has yet been found. Since there are telescopic images taken of positions that the parent must have occupied (determined by tracing back the orbit of the fragment chain), the parent must have been too faint to show up on any of them. Adding an assumption about the inherent brightness of the comet gives an upper limit on its diameter of about 10 km. Naturally, the fragments in the chain must be a fraction of this size.

Another estimate of the fragments' sizes came from the HST images themselves. Although the HST could not resolve the nuclei directly, the light in the brightest pixel in the image of a fragment is composed of sunlight that is either reflected from the nucleus itself or from the surrounding dusty coma. If the amount of light reflected by the coma alone can be determined, then what is left over must come from the nucleus. Knowing the amount of light reflected by the nucleus, and adding the usual assumption about its intrinsic brightness (usually that SL9 was no brighter than the very dark Halley's Comet), the nucleus size can be extracted.

The trick is to deduce how much light comes from the coma alone. This was done by measuring the brightness of pixels adjacent to the brightest one. Since these contain light only from the coma, the coma's brightness some distance away from the nucleus can be determined. By looking at pixels even further out, a general rule for how the brightness of the coma depends on distance from the nucleus can be constructed. Extrapolation of this rule backward to the brightest pixel then reveals how much light the coma contributes to that pixel. Simple

subtraction should then determine the amount of light contributed by the nucleus itself. Unfortunately, the relation between the coma's brightness and distance from SL9's nuclei is very different from that of other comets, leading to some uncertainty in the validity of this procedure. Also, the extrapolation showed that most of the light in the central pixel of each of the SL9 fragments came from the coma, not from the nucleus, so the HST astronomers were in the uncomfortable position of subtracting a large but uncertain number from the observed brightness. The best estimates made in this way suggested that the largest fragment could have been as much as 4 km in diameter, although as time went on, most of the HST astronomers began to lose confidence in this estimate and emphasized that it was very much an upper limit. They could not be certain that the nucleus was contributing any light to the brightest pixel!

Another indirect estimate of SL9's size came from the way it broke up during its first close approach to Jupiter in 1992. The fact that it broke up at all is somewhat

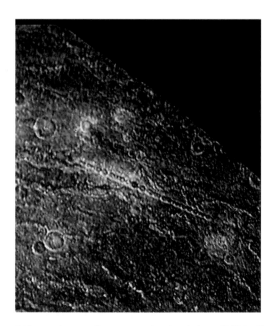

A *Voyager* image of an unnamed crater chain on Callisto, the outermost Galilean moon. It is 360 kilometers long and contains craters up to 24 kilometers across. Such chains were a mystery till the discovery of SL9, when Jay Melosh and Paul Schenk realized that if a comet broke up as it passed close to Jupiter, as SL9 did, and then ran into one of the Galilean satellites, it would produce a crater chain just like this. The idea is strengthened by the fact that all the crater chains on Ganymede and Callisto are on the hemispheres facing Jupiter. (JPL/NASA; processing by P. Schenk)

astonishing. Even though Jupiter has a powerful gravity field, the tidal stresses – which are the only forces that can tear apart a passing comet or asteroid – are unbelievably weak. If SL9 had a strength of only 0.001 bar – far weaker than the fluffiest soufflé any cook on Earth could prepare – it would not have broken into fragments. In its original form, SL9 must have been a very fragile body, perhaps a composite of smaller subunits held together only by the feeble forces of their own mutual gravitational attraction.

There is a further possibility. Analysis of SL9's orbit showed that it passed through the plane of Jupiter's faint ring. The comet might have collided with a ring particle and have been shattered by the impact. In either event, however, the parent comet broke into fragments during a very close pass by Jupiter.

Since all the fragments probably started out traveling with very nearly the same velocity, their subsequent paths were determined by their relative positions at the time of break-up. The length of the observed chain of nuclei at any later time gives the separation of the fragments at the moment of break-up and so the parent comet's diameter. Jim Scotti and I first made this computation in June 1993 and found a surprisingly small diameter of 2 km for the parent. Several other people have subsequently made similar computations, using increasingly sophisticated models for break-up, but most get basically the same result. This model implies that the largest fragment was probably not more than a kilometer in diameter, and that most of the fragments were considerably less than that.

An even more indirect estimate of the size of SL9 comes from the crater chains observed on Jupiter's satellites Ganymede and Callisto. Although it may seem that such crater chains have little to do with SL9, it now looks likely that these chains record the past impacts of other split comets. They may be able to tell us about the characteristics of the comet population of which SL9 was a member. The origin of these enigmatic chains was not understood until Paul Schenk and I were inspired by seeing Jim Scotti's first image of the 'string of pearls' in the newspaper. We both independently recognized that if SL9 had happened to run into one of Jupiter's moons instead of flying (temporarily) out of the jovian system, the result would have been a chain of craters just like those observed. Subsequent analysis showed that the fragments that created these chains were typically a kilometer in diameter or smaller, consistent with the size suggested by the tidal break-up models. Of course, this result, based on about a dozen crater chains, does not prove that SL9 had to be about the same size – it could have been a giant among its brethren – but the likelihood of such a situation is small.

So, now that the impacts have occurred and have been observed by the largest array of astronomical power ever trained on a single event, what is the answer? How big were the fragments and, by addition, the parent body of SL9? Unfortunately, we still do not know! The phenomena of the plumes are still not well enough understood to make a firm estimate of the total energy released in each impact. Eventually, a consensus will probably be reached, but not before many more months of analysis.

Three examples of crater chains on Ganymede. Crater chains on Ganymede tend to be shorter and less conspicuous than those on Callisto, perhaps because any impacting comet would have had less time to spread out before running into Ganymede, which is closer to Jupiter than Callisto. Ganymede's surface is also younger, so it has accumulated fewer crater chains. The central panel shows what looks like a single elongated crater, but the line of multiple central peaks reveals that it was produced by the near-simultaneous collision of a line of impactors. (JPL/NASA; processing by P. Schenk)

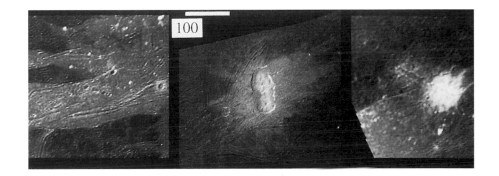

The play of the plumes

One of the most spectacular of the phenomena accompanying the impacts were the plumes kicked up by the fragments as they plummeted into Jupiter's atmosphere at 60 km/s. In the wake of six months of analysis, a general consensus has nearly emerged about how the plumes formed. This is partly because many of the plume phenomena were anticipated by the theorists before the impact (see Chapter 5), but there were also many surprises!

Observers have now recognized three main phases of plume formation. There was an early, bright flash, the spectrum of which indicated very high temperatures. It corresponded either to the entry of the bolide into the atmosphere, or to the earliest stages of the expanding fireball after the fragment had deposited its energy in Jupiter's upper atmosphere (or maybe both). The second phase was seen by observers on Earth as a dusty plume that rose above Jupiter's limb. This was an expanding cloud of gas and debris that rose up 3000 km. The final phase was observed in the infrared, when the gas and debris re-impacted Jupiter over a region 10 000 km across and converted a large fraction of the impact energy to heat.

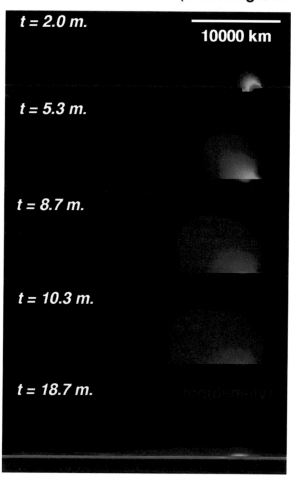

Hubble Image of Impact G

7:33 UT   M

7:38 UT   R

7:41 UT   G

7:44 UT   B

7:51 UT   V

July 18, 1994

3-D Fireball Simulation (3-km fragment)

t = 2.0 m.   10000 km

t = 5.3 m.

t = 8.7 m.

t = 10.3 m.

t = 18.7 m.

Comparison of the HST image sequence of the G impact plume rising over the Jupiter limb with a simulation by the group at Sandia National Laboratories shown at approximately the same scale and with similar timing. The G impact occurred at 7:33:32, at about the same time as the first HST frame: the first simulation image is from 2 minutes after impact. The other simulation images correspond more closely in time with the HST image to their left. The portion of the simulated fireball that would be below the jovian horizon is not shown, to aid comparison with the HST images. Note the tilted glowing column below the sunlit part of the impact plume in the 7:38 UT HST image: this corresponds nicely with the tilt of the simulated fireball, which escapes to along the comet's entry trajectory. The small bright spot to the lower left of the impact plume in the 7:41 UT image is an artifact of the HST camera. (M. Boslough, Sandia National Laboratories)

The first two phases of plume development were more or less anticipated by the theorists, but the splashback was a surprise to nearly everyone. In hindsight, of course, we should have expected it. Several years before the discovery of SL9 Kevin Zahnle and I (along with Nick Schneider and Don Latham) argued that the heat liberated by impact ejecta falling into Earth's atmosphere was responsible for igniting wildfires that raged over the entire Earth 65 million years ago following the impact that wiped out the dinosaurs. We should thus have anticipated a similar phenomenon on Jupiter, but we did not. Nevertheless, it is pleasing to recognize on Jupiter what we thought occurred on Earth. The observation of the splashback phenomenon on Jupiter also greatly strengthens our confidence in the effect and has raised further everyone's awareness of the danger that impacts pose to the Earth. Not only is it dangerous to be in the near vicinity of an impact but the ejecta also pose a hazard, and over a much broader region, as we saw on Jupiter.

Deep or shallow penetration?

Another vexing question, still not completely resolved, is 'How deep did the comet fragments penetrate?' The answer to this question is partly tied to the other unresolved question of the fragment sizes. Many of the pre-impact computer simulations assumed the fragments were a kilometer or more in diameter, and thus predicted the bolides descending hundreds of kilometers below the visible cloud decks. The observations, however, suggest much shallower penetration. Most of the evidence comes from the detection of different chemical substances in the plumes.

Jupiter's cloud decks are believed to be arranged in three distinct layers. The uppermost layer (which can be observed directly) is formed of ammonia crystals, NH_3. A deeper (and hypothetical) layer is made of ammonium hydrosulfide crystals, NH_4SH, while the deepest is made of either ice or liquid water drops, H_2O (see Chapter 4). Spectra taken by the HST of the early plumes show evidence for sulfur that may have been excavated from the jovian clouds. If so, the fragments reached the ammonium hydrosulfide cloud deck at about 30 km below the visible cloud top and that the ensuing fireballs entrained a large quantity of sulfur from Jupiter itself. Although water was (finally) detected in the spectra, its abundance was so low that it could all have originated in the comet fragments themselves. In addition, metals such as iron and magnesium were detected. Most likely, these also originated in the comet fragments. These observations argue that the fragments did not reach the water clouds 60 km below the visible cloud tops, and so the ultimate penetration was shallow.

What could have caused the penetration to be so shallow? Size is one important factor, but the fragments' strength properties also played a role. If the cometary fragments were small (less than a kilometer in diameter) and broke up easily under aerodynamic forces during their entry, they would have spread out, increasing the area they presented to the atmospheric gases streaming by and would have slowed down much more quickly than if they had remained in a solid chunk. This would lead to their stopping and exploding high in the atmosphere, and perhaps can explain why no jovian water was entrained in the plumes. If the comet fragments had been larger, many kilometers in diameter, even atmospheric break-up could not have halted their penetration through the water clouds.

Another possibility is that the comet nuclei were not solid, but instead loose swarms of boulder-size chunks of ice and rock before they struck the atmosphere. A projectile like this, even if it had the mass of a multi-kilometer solid body, would also be braked quickly in the atmosphere. (Compare the drag of a parachute packed in its bag with one that is open.) A few lines of evidence still argue for penetration to, or below, the water clouds. Andy Ingersoll and his colleagues believe that the waves that were observed propagating away from the impact sites were trapped in the jovian troposphere and that to generate such waves the nuclei must have penetrated at least this deep. Numerical simulations by Marc Boslough and his colleagues have shown that the plume phenomena may be nearly independent of the size of the projectile. A large object may simply leave behind a small fraction of its total energy in the upper atmosphere while the main body carries on deep into the atmosphere without generating a visible fireball.

What is that dark stuff anyway?

For amateur astronomers the most visible effects of the comet impacts were the immense, Earth-sized dark patches that look rather like bruises on the southern hemisphere of Jupiter. They were evidently some kind of fine dust floating in Jupiter's atmosphere above the ammonia clouds. Since they had not settled out six months after the impact, they must have been extremely small particles, less than a few microns in size. But what is the dust? Some of it may be condensed silicates and metal oxides from the comet fragments themselves, but there is another intriguing possibility. Some of this material may be organic mixtures ('tholins') synthesized in the comet fireballs.

Jupiter's upper atmosphere is a huge chemical factory rife with possibilities for the synthesis of organic materials. Experiments to investigate the possible origin of life on Earth have involved electric sparks in a gaseous mixture of ammonia and water to produce amino acids and other primitive precursors of life. Although it is now doubted that Earth's atmosphere ever resembled this mixture, it is similar to the present atmosphere of Jupiter. Carl Sagan suggested many years ago that the red color of Jupiter is due to organic molecules synthesized, not by electric sparks (although lightning does occur there), but by the action of solar ultraviolet radiation on the atmospheric gases. (Others believe, however, that sulfur or phosphorous are responsible for Jupiter's colors.)

The sizzling fireball created by a comet fragment impacting at 60 km/s is just the place for a lot of chemical synthesis. Initially, the fireball is so hot that all the molecules in it are broken up. Later, however, the fireball cools, entrains some surrounding atmosphere, and many different molecules can assemble themselves. Unlike when chemicals react in a jar on Earth, equilibrium was never reached in the expanding fireball, and many exotic molecules may have been showered over a broad area of Jupiter's surface as the plumes fell back. If the dark spots did have a significant content of organic material, we may have been witness to a process similar to the one that, some four billion years ago, created the organic precursor molecules out of which life arose on Earth. There has been much speculation about the role of impacts in the origin of life, and SL9 may have slotted another piece into the puzzle. Certainly, the topic of chemistry in the expanding fireball of an impact is something that will get a lot more attention in the future.

Aurorae and radio waves

Before the impact of SL9 very little thought had been given to the electromagnetic effects of an impact. Although the generation of an 'electromagnetic pulse' is a well-known side effect of a nuclear explosion, such phenomena are short lived and leave no known record of their existence, so little work had been done to analyze electromagnetic disturbances caused by an impact. Nevertheless, it is clear that the fireball generated in an impact is so hot that the gases will be strongly ionized. Ionized gas moving rapidly in a magnetic field can generate electric fields. It was thus gratifying to find that the impacts did have some effects on Jupiter's magnetosphere, although they are not yet completely explained.

The most spectacular electromagnetic effect of the impacts was the appearance of bright aurorae in the *northern* hemisphere of Jupiter at places connected to the impact sites in the southern hemisphere by magnetic field lines. These aurorae may have been caused by electrons and ions from the fireball traveling along the field lines. Similarly, radio emission from the jovian magnetosphere intensified in the wake of the impacts. Analysis of these phenomena is not yet complete, but eventually we may possess a better understanding of the electromagnetic effects of impacts on Jupiter, in general, and especially on Earth.

Summing it up

So what, in a nutshell, have we learned from the impacts of SL9 on Jupiter? I personally believe we are wiser at two different levels. At the technical level, we have learned a great deal about the short term phenomena that accompany an impact. Although craters on the surfaces of the rocky planets record the long term effects of an impact, the phenomena of projectile fragmentation on entry, fireball formation and the spectacular plume, were all stuff of speculative theory beforehand. In retrospect, however, none of the observed phenomena were a complete surprise and we believe that they can all be explained by appropriate extensions of existing models. The impact of SL9 will not initiate a 'paradigm shift' in the science of impact cratering, although many developments of existing models can be expected.

On a more emotional level, we have seen directly that large celestial impacts do occur and could be devastating to civilization on Earth. In the long run, this realization may be the most significant outcome.

10

What If?...

Clark R. Chapman

Could a cosmic impact cause devastation on Earth in the future? The sight of huge black spots on Jupiter following the impact of Comet Shoemaker–Levy 9 convinced many people that the threat should be treated seriously, even if such events are very rare.

For millenia, comets were generally feared. What were those fuzzy interlopers among the fixed stars? People took them to be harbingers of doom. A modern, scientific perspective on the nature of comets and asteroids has developed during the twentieth century. So has a realistic appreciation of their potential impact – literally and figuratively – on our planet Earth. Our ancestors' fears were not without merit.

Impacts, craters and catastrophes

Soon after the discovery of the first Earth-crossing asteroid, several astronomers made the connection with the cratered face of the Moon. In his 1949 book, *Face of the Moon*, Ralph Baldwin wrote that if the fresh lunar crater Tycho had instead formed on our world today, it would be 'a horrifying thing'. In the late 1950s, a youthful Eugene Shoemaker discovered coesite – a high-pressure mineral created by impact explosions – at Arizona's Meteor Crater, clinching its reputation as an extraterrestrial creation just 50 000 years old. The late Irish-Estonian astronomer Ernst Öpik calculated orbits of Earth-crossing asteroids and comets; he realized that planetary surfaces must be cratered. The early *Mariner* probes to Mars and Mercury soon confirmed his prediction.

In 1980, the late Nobel laureate Luis Alvarez, his son Walter, and their Berkeley research team published their idea that a comet or asteroid caused one of the greatest catastrophes recorded in the geological record – the mass extinctions which took place at the transition between the Cretaceous and Tertiary eras of geologic history. Debate during the 1980s centered on what had happened in Earth's geological past, not what might happen today. Until the last few years, only science fiction writers like Arthur C. Clarke, Jerry Pournelle, and Larry Niven had the imagination to slam a comet into Earth and speculate about the consequences for civilization.

In 1989 several events conspired to bring the impact hazard to the attention of the scientific community and the public. In our 1989 book *Cosmic Catastrophes* David Morrison and I reported on an obscure 1981 meeting, chaired by Eugene Shoemaker, at which participants analyzed the impact hazard and established much of our basic understanding of the threat. A related seminar, given by Morrison to the US Congressional Space Caucus may have been the first time that the issue was presented to policy makers.

A real or imagined threat?

In March 1989, a 300-meter-sized object hurtled within 700 000 km of Earth. It was just one of the ever increasing numbers of comets and asteroids being discovered by the patient telescopic search teams of Eleanor Helin, Tom Gehrels, Carolyn and Eugene Shoemaker, and their colleagues. But an exaggerated – actually erroneous – press release by NASA placed the

Meteor Crater near Flagstaff, Arizona, is 1.2 km in diameter and 200 m deep. It was excavated about 50 000 years ago by the impact of an iron asteroid about 40–50 m across and is one of the youngest impact craters on Earth. (D. Roddy, USGS)

'what-if-it-had-hit' issue on the front pages of newspapers around the world.

As the Cold War wound down, American aerospace engineers looked for new applications of their 'Star Wars' detection and interception technologies, and began pressing the government to take action against a new enemy: asteroids and comets. They convinced the House Committee on Science, Space, and Technology to ask NASA to study the impact hazard. They also asked the then Vice President, Dan Quayle, to speak on the subject. That was a mistake. Quayle was habitually ridiculed by the press, and a talk he gave to the American Institute of Aeronautics and Astronautics in June 1990 only cemented his reputation as a laughingstock. When the NASA Committees gave their recommendations to Congress in early 1993, the subject's 'giggle factor' kept it on the back burner. 'Now we also have to worry about getting hit on the head by a comet? Get real!'

Comet Shoemaker–Levy 9 killed off the giggle factor. It was not really the comet, or even the improbability of its striking Jupiter. Nor was it the spectacular fireworks of Comet Crash Week, which testified as much to the sensitivity of modern astronomical instruments as to the energy of the impacts. Indeed, the impact flashes, measured directly by the Galileo spacecraft, were only a few percent the brightness of Jupiter itself, much dimmer than many had predicted. What finally made the

The big, black spots

Within 24 hours of the initial impact, professional astronomers reported on the electronic mail 'exploder' that new dark spots were visible on Jupiter, at the impact sites of fragments A, C and E. Some spots were even said to be readily seen through the finder telescopes and on screens of the video guiders. I began to realize what was happening on 17 July. The previous night, I had been recording CCD data with the 2.1-m telescope on Kitt Peak, 60 miles from my home in Arizona, hoping to detect the impact of fragment B. (Unhappily, it was a dud.) But the powers-that-be at the National Observatory on Kitt Peak felt the comet crash was not important enough to keep the telescopes open, so the telescope was scheduled to be shut down for the rest of the week. I was left to observe Jupiter with my 10-inch (25-cm) backyard telescope, which I had used as a high-school boy to study Jupiter's cloud motions.

That Sunday evening, as I glimpsed the new black spots rotating across Jupiter's southern hemisphere, I understood how inadequate our predictions had been. I had written in the magazine *Sky & Telescope* that a small spot might be visible within a couple hours of impact, and that 'a bright new layer of stratospheric haze may form in far-southerly latitudes and become prominent through backyard telescopes' during the week of impacts. My colleagues thought I was too optimistic feared the whole SL9 event would be a 'cosmic fizzle'. That evening, my wife, my daughter, and my daughter's friend all saw the new spots through my telescope, despite poor seeing conditions. Evidently, I hadn't been optimistic enough. And, as I suppose we should have realized, the spots were *black*, not white.

The following evening was a shocker. Kitt Peak was still closed, though the media excitement finally convinced the Acting Director to agree to reopen it for fragment V's impact at the end of Comet Crash Week. I peered through my 10-inch again, and saw the aftermath of fragment G's demise. Its towering plume had been captured by the Hubble Space Telescope less than 24 hours before. I was stunned. Not only was G's site the largest, darkest spot I had ever viewed on Jupiter, it was more prominent than any recorded since telescopes were invented – and I had pored over most published observations from Cassini's time (*c.* 1665) through the British Astronomical Association's invaluable records of the last century. I wrote of G's unprecedented nature on the Internet, which helped spur awareness among

The great black spots appeared practically at the same time as the growth and collapse of the fireballs. Several observers reported that they appeared to get even darker in the first day or so of their existence, perhaps as the site cooled and more solid particles condensed. This image, taken at 20:30 UT on 18 July with the 1-m Jacobus Kapteyn Telescope on La Palma in the Canary Islands, shows the G impact spot on its second rotation around Jupiter. Ganymede can be seen in the upper left of Jupiter's disk, beginning a transit across the planet. (Courtesy Alan Fitzsimmons)

amateur astronomers, science educators, and the public at large that *they* could participate in the SL9 drama themselves. The smallest telescopes were sufficient to track the larger impact scars during impact week and through August. Then Jupiter's 1994 period of visibility from Earth drew to a close, and it became lost in the Sun's glare.

No longer a joke

Jupiter's largest bruises – the zones of destruction – greatly exceeded the size of our own planet Earth. Hundreds of thousands of people could see with their own eyes the havoc wrought by the impact of just one piece of a modest-sized comet. The Earth's orbit is crossed by thousands of asteroids and comets larger than the G fragment. So the notion that the threat from cosmic impacts did not deserve serious attention was finally put to rest. Before the last fragment plunged into Jupiter, the US Congress swung into action and asked NASA to formulate plans to deal with the impact hazard. A few days later, NASA appointed none other than Eugene Shoemaker to head the committee. He was charged with determining how the government would 'identify and catalogue within 10 years the orbital characteristics of all comets and asteroids that are greater than 1 km in diameter and are in an orbit around the Sun that crosses the orbit of Earth'.

What do the bruises on Jupiter signify? Is the danger even worse than we thought, or are our fears exaggerated? Are the black spots nothing more than superficial, transient stains on Jupiter's appearance? Or do they mean that Earth would be darkened from a comet impact, too? How long would the effects last and how hazardous would they be? What can we learn about the impact hazard from what happened on Jupiter in July 1994?

We don't know all the answers yet, but one thing is clear: Shoemaker's 1981 workshop, as verified by more rigorous studies after the 1989 'near miss', had it right. Though the danger to Earth is not much worse than experts had thought, there are fewer doubts, and the public finally has good reason to take the hazard seriously.

Potential for catastrophe

The Earth is struck, on average, every few hundred thousand years by objects a mile across or larger. That is roughly the minimum size of impactor that can do global

Earth one and three-quarter hours after the impact of the G fragment, had it hit our planet instead of Jupiter. The approximate appearance of the impact scar in the HST images has been 'wrapped' onto Earth at the correct scale, so that distances of features from the impact site in this image are the same as they were on Jupiter. The impact occurred near Detroit, blackening skies over the entire eastern US, and the dark atmospheric wave is speeding across the Atlantic towards Europe, Africa and South America. Dark material from the collapsing fireball blankets southern South America and the South Atlantic. Of course, this simulation assumes that the processes that created the black clouds on Jupiter would work the same way on Earth, and this is debatable, but after the SL9 impacts there can be little doubt that similar-sized impacts on Earth would have global consequences on a scale similar to that pictured here. (J. Spencer)

damage sufficient to threaten the future of civilization as we know it. It takes an even larger, rarer asteroid or comet to create a holocaust that would render the human species and others extinct, and manifest itself as a mass extinction in the fossil record. On the other hand, objects less than a mile across strike more frequently and could do terrible, but localized damage. Even the 50-meter object that levelled more than a thousand square kilometers of Siberian forest in 1908 and killed practically nobody could have exterminated millions had it hit a major city. But a hundred times as many major disasters are caused by natural hazards, like floods and earthquakes, as by cosmic projectiles less than a few hundred meters in size.

The G fragment struck Jupiter with roughly the energy of the mile-wide objects that are the greatest threat to civilization. Its impact is as good a picture as we have of what could happen to us. So what happened on Jupiter?

The *Galileo* spacecraft measured the color temperatures of the brilliant, luminous phases of the G impact during the first seconds and minutes. A brilliant blue-white meteor or bolide flashed through Jupiter's skies for a few seconds. Any cloud-top jovian creatures would have seen a streaking flare a hundred times brighter than the Sun, as the bolide soared to 10 or 20 thousand degrees K. As the disintegrating fragment plunged beneath the clouds, the superheated atmosphere above formed a fireball about 10 km across. Over the next minute or two, the expanding bubble of gas grew hundreds of times in volume as it cooled and began to erupt explosively into space, according to *Galileo*'s infrared spectral data.

Hubble's camera captured what happened during the next 10 minutes, as the plume of gas – contaminated, of course, by material ablated from the G fragment during its fiery entry – towered more than 3000 km above Jupiter's cloud deck. The arching plume began then to cascade back down across an enormous region of Jupiter more than 20 000 km across – perhaps twice the dimensions of Earth. Re-impacting the top of Jupiter's stratosphere at roughly 10 km/s, the glowing secondary debris heated much of the impact region to nearly 1000 degrees. *Galileo* saw the heat beginning just 6 minutes after impact, and ground-based astronomers marvelled as the firestorm reached a crescendo 10 minutes later as Jupiter's rotation carried the hell hole into direct view from Earth. Any jovian creature in the open, beneath the firestorm on Jupiter, would have been fried.

Temperatures declined over the next few hours, leaving the infamous black pall. Meanwhile, a mighty wave swept across the impact region, reverberating from the final demise of the comet fragment deep within Jupiter's atmosphere, far below the clouds. Much of the black material was made of tiny, micron-sized aerosol particles, floating in rarefied gas at the top of Jupiter's stratosphere. Larger particles were soon settling down toward the clouds. Over the ensuing weeks and months, upper atmospheric winds distorted the shape of G's graveyard, stretching it out to merge with other impact bruises, forming a black belt more than 100 000 km long.

One might imagine that the G's black pall would have shielded the surface below from the Sun, putting any local jovians who had escaped the firestorm into a dark deep freeze. Actually, calculations by Richard West show that sunlight is dimmed only slightly beneath the black haze. Yet, on Earth, lowering surface temperatures by just 10% is the difference between summer and winter – nothing to scoff at. More significant, according to West, would be stratospheric heating and resulting changes to upper atmospheric chemistry and winds. As the ozone hole has recently reminded us, we care a lot about what happens to our own planet's stratosphere, even though we don't live up there.

If it happened here

Astronomers may debate for some time whether the black material was mostly derived from the comet or from Jupiter's atmosphere. Comets (and most asteroids) have abundant carbonaceous materials. To whatever degree the disintegration and high-temperature processing of the comet yielded the black palls on Jupiter, we might expect something similar if a comet struck Earth. However, if the aerosols were primarily high-temperature products of Jupiter's atmosphere – e.g. from the carbon-bearing methane that constitutes a fraction of a percent of Jupiter's air – then Earth's chemistry might yield different stratospheric aerosols, which could be less harmful, or

In mid December 1994, Jupiter emerged into the pre-dawn sky after being out of view behind the Sun. The first observations of the impact sites were reported by David Gray, a British amateur who observed Jupiter visually on 14 December and noted that the impact latitudes were still dark, a reminder that a similar impact on Earth could have long-term consequences. Comet co-discoverer David Levy confirmed this sighting on 18 December. Both observers noted that the impact latitudes were now only the second most conspicuous feature on Jupiter, after the North Equatorial Belt. Seeing conditions were poor for David Gray in mid-December, but improved by the 26th when he made the drawing on the left. The drawing on the right was made three months later on 12 March 1995. The dark belt is fading and breaking up, but still present. (David Gray)

The long life of the impact clouds is demonstrated by these infrared images (1.64, 2.12 and 2.26–2.29 microns, left to right) taken at 13:45 UT on 14 January 1995 with the 2.4-m Hiltner Telescope of the Michigan–Dartmouth–MIT Observatory on Kitt Peak. In the following months Jupiter's appearance slowly returned to normal, but we know that huge impact clouds will some day blossom there again, and on other planets, including our own. It is only a matter of time. (M. Skrutskie, D. Morris & K. Allen)

more so. So far, calculations for dangerous impacts on Earth have assumed a global distribution of silicate particulates, blown up from the impact crater, which shield sunlight less effectively than the sooty particles formed by SL9 on Jupiter. This is one lesson from the Jupiter crash that should, perhaps, make us worry more.

Another lesson concerns the firestorm. The plumes soared higher than predicted, and the subsequent, fiery re-impact of plume debris across a billion square kilometers of jovian territory was, therefore, more energetic than had been thought. Calculations by O. B. Toon and colleagues for a terrestrial impact have suggested that lethal global heating might require a comet much larger than SL9, but maybe smaller impactors are more dangerous than we thought. Surely, if an impact is ever predicted, it would be good to be indoors, preferably underground, during the hours following impact – even if the point of impact is in another continent.

So far, SL9 has not changed our estimates of how frequently comets and asteroids strike Earth. But the comet's tidal break-up highlights how little we know about the smaller bodies in the solar system. Eugene Shoemaker now believes that many Earth-threatening comets might actually be fragments of precursor comets that pass too close to Jupiter and break up, like SL9. Such fragments would probably be sprayed throughout the inner solar system. (SL9's dive back into Jupiter was something of a fluke.) As we learn more about the nature and behavior of asteroids and comets, our perceptions about the impact hazard will surely change.

An over-riding lesson from the Jupiter crash is that our knowledge of cosmic processes is woefully inadequate. Some of those who were bold enough to announce predictions before the impacts can take comfort in the fact that they got some things right. But other elements crucial for evaluating the terrestrial impact hazard were missed. We needed to observe Nature's experiment to see what really happens.

Wisdom is realizing how little we understand. With the future of civilization hanging in the balance, however small the odds, it behoves us to learn more about the impact hazard. I hope that, in the years ahead, the mountain of comet crash data obtained will be thoroughly analyzed and synthesized with knowledge about Earth's atmospheric chemistry, about hypervelocity impacts, and about comets and asteroids. We may then be in a better position to understand how seriously to take the threat from the skies. Governments are often short-sighted in supporting the basic research required to understand the relationship of human beings to our planetary environment, and Earth's relationship to the cosmos. There will be pressures to throw billions into military hardware capable of blowing up a cosmic projectile that will almost certainly never threaten us in our lifetime. Yet a tiny fraction of such expenditure could build the telescopes that would prove whether or not we are in danger – whether another, as yet undiscovered comet is heading toward Earth. Such a search program would almost certainly provide the necessary lead time to mount a deflection mission, or take other steps to mitigate disaster, before it struck. And just a fraction of the funds required for a Spaceguard Survey could support the research about comets and asteroids and about Earth's atmospheric and biospheric processes that could teach us how vulnerable we are to what we witnessed on Jupiter.

Glossary and Abbreviations

AAT Anglo-Australian Telescope

accretion The process by which small particles or objects come together under their mutual gravitational force to create a larger mass, or the accumulation of material on an existing larger object.

aerosol A suspension of solid or liquid particles in a gas.

ANU Australian National University – It operates observatories at Mount Stromlo near Canberrra and at Siding Spring in New South Wales.

arcminute One-sixtieth of a degree, a measure of apparent size in the sky. The full Moon is about 30 arcminutes across.

arcsecond One-sixtieth of an arcminute. Jupiter was about 40 arcseconds across at the time of the impacts.

asteroid A small rocky body in the solar system. Several thousand are individually known, all with diameters of 1000 km or less.

AU Astronomical Unit. A unit of measurement for distances in the solar system equivalent to 149 597 870 km (92 955 730 miles), which is derived from the mean distance between the Earth and the Sun.

aurora A luminous emission in the atmosphere of a planet, caused when gas atoms are excited by collisions with energetic ions or electrons.

bar A unit of measurement for pressure, particularly that of a planetary atmosphere. One bar is close to average pressure of the Earth's atmosphere at sea level and is equivalent to 100 000 pascals (newtons per square meter). 1 millibar (mbar) is one thousandth of a bar.

bolide A particularly bright or explosive meteor.

CCD Charge-Coupled Device. An electronic imaging device composed of a matrix of light-sensitive pixels. CCDs are sensitive in the wavelength range covering the ultraviolet, visible light and the near infrared to 1 micron.

Centaurs Asteroids with orbits between Jupiter and Neptune. The prototype is Chiron, named after a centaur appearing in Greek mythology.

CHON particles Particles found in comets, composed of light elements and rich in carbon (C), hydrogen (H), oxygen (O) and nitrogen (N).

coma The diffuse envelope of gas and dust surrounding the nucleus of a comet.

conjunction An alignment of two astronomical bodies so that they appear to be at the same place (or close) in the sky.

corona The faint outermost part of the Sun, normally visible only during a total eclipse of the Sun.

dust tail One of the two types of tail comets develop as they approach the Sun. Dust tails are composed of particles about one micron in size which are driven off by solar radiation pressure.

Earth-crossing asteroid An asteroid in an orbit which crosses the orbit of Earth. They include members of the Apollo and Aten groups.

eccentricity One of the parameters used in the mathematical description of an ellipse, which is the basic path followed by an object in orbit. For an ellipse, the greater its eccentricity on a scale from 0 to 1, the greater its deviation from a circle.

ecliptic The plane of Earth's orbit around the Sun.

ecliptic north The northerly direction perpendicular to the plane of the ecliptic (as opposed to the plane of Earth's equator).

ejecta Material excavated by an impact or thrown out from a volcano.

emission line A bright band or peak in a spectrum produced by light of a specific wavelength emitted by a gas. The wavelength is diagnostic of the composition of the gas.

ephemeris (plural ephemerides) A prediction of the location of a celestial object as a function of time.

ESO European Southern Observatory – a European astronomical research organization, with headquarters in Germany and observatories in Chile.

exploder A limited electronic mail communication network which automatically distributes messages posted by one member to all the others.

Galileo A NASA spacecraft mission to Jupiter, launched in 1989 for arrival in 1995. En route, it returned images of two asteroids, Gaspra and Ida, but has been hampered by the failure of its high-gain communications antenna.

Hipparcos A European Space Agency satellite, launched in 1989, to measure the positions, distances and movements of stars.

HST Hubble Space Telescope – a 2.4-m orbiting telescope operated by NASA and ESA

hypersensitize To treat a photographic film or plate, chemically or by heating, so as to increase its sensitivity to light. The emulsion deteriorates relatively quickly after such treatment.

IAU International Astronomical Union – the international cooperative organization for astronomical affairs.

infrared Electromagnetic radiation with wavelengths longer than those of visible red light. The wavelength range described as infrared is from about 0.7 to 100 microns.

ions Electrically charged atoms. Atoms have a net positive charge when they lose one or more of their electrons.

IRTF Infrared Telescope Facility – a NASA infrared telescope at Mauna Kea Observatory, Hawaii, operated as a US national facility.

JPL Jet Propulsion Laboratory – an institution at the California Institute of Technology, Pasadena, operated in support of the programs of NASA and other agencies.

Kuiper belt A population of icy bodies in the solar system occupying a disk-shaped region beyond the orbit of Neptune and extending outwards perhaps for thousands of light years. It is thought to be a source of comets.

light curve The variation in brightness of an object over time, usually represented as a graph.

limb The extreme edge of the visible disk of a planet, moon or other astronomical object.

Magellan A NASA spacecraft, launched in 1989, which was placed in orbit around Venus and mapped the surface of the planet by means of radar.

magnetosphere The region around a planet that is dominated by the planet's own magnetic field and constrained by the flow of the solar wind around it.

magnitude A measurement of the brightness of a celestial object on a scale in which the lowest numbers describe the brightest objects. A magnitude difference of 5 corresponds to a brightness ratio of 100.

Mariner A series of spacecraft launched by NASA in the 1960s and 1970s to explore the planets Mercury, Venus and Mars. Mariner 4, launched in 1964, was the first successful probe to Mars, revealing impact craters on the planet's ruface. Mariner 10 returned the only close-up images ever secured of Mercury in 1974 and 1975.

mass The amount of matter contained in an object, as distinct from its weight, which is a measure of the gravitational pull exerted on the object and varies with the local strength of gravity.

mbar see bar

meteoroid A piece of rock or dust in space.

micron A unit of measurement used, for example, in describing the wavelength of infrared radiation and the size of dust particles. It is a popular alternative for the more formal name of this unit, the micrometer (μm) and is one millionth of one meter.

nanometre A unit of measurement used particularly to describe the wavelength of visible and ultraviolet light. One nanometer is one billionth of a meter.

NEO Near Earth object – an asteroid or meteoroid which passes unusually close to Earth.

nm Abbreviation for nanometer.

NRAO National Radio Astronomy Observatory – the national facility for radio astronomy in the USA, which operates telescopes at various locations, including the Very Large Array (VLA) in New Mexico.

NSF National Science Foundation – a US science funding agency.

Oort cloud A cloud of billions of icy bodies, which are potential comets, thought to surround the solar system in a spherical shell at a distance of about 1 light year (50 000 light years).

parabolic orbit A planetary orbit which is so elongated that its furthest point from the Sun is effectively at infinity.

perijove For an object orbiting Jupiter, its closest approach to the planet.

photometer An instrument for measuring accurately the brightness of an object.

pixels Short for picture elements – the smallest individual elements in an image obtained electronically, for example with a CCD.

plume An informal term for a mass of material ejected upwards from a point source, such as an impact or volcano.

radiation pressure The physical pressure exerted by light, or other electromagnetic radiation, which is strong enough to blow a dust tail off a comet, for example.

R_J The standard abbreviation for a distance equal to the equatorial radius of Jupiter, i.e. 71 492 km.

Roche limit The minimum distance from the center of a planet that a weak satellite can orbit and remain stable against destruction by tidal forces.

Schmidt telescope A type of wide-field astronomical telescope, often used for survey work.

spectrum The result of dispersing a beam of electromagnetic radiation so the components of different wavelength are separated.

stratosphere An upper region of a planetary atmosphere, overlying the troposphere. The stratosphere is stable layer in which convection does not take place.

STScI Space Telescope Science Institute – a research institute in Baltimore, Maryland, operated under contract to NASA to manage the science programme of the Hubble Space Telescope.

sublimation The process in which a solid material changes directly into a vapor or gas without a liquid phase in between.

synchrotron radiation Electromagnetic radiation emitted by electrically charged particles travelling at high speed through a magnetic field.

System II, System III Systems for specifying longitudes on Jupiter, which presents problems because of the lack of a solid reference surface. System II is based on a rotation period of 9 hours 55 minutes 40.6344 seconds, derived from observations of features near the Great Red Spot, and is generally used by amateur observers. System III is the rotation rate of Jupiter's interior. System III and its adopted period of 9 hours 55 minutes 30.11 seconds is system used in most professional studies. Longitudes in the two systems at a time t are linked by the equation:

$L_{III} = L_{II} + 81.245 + 0.266 (t - t_0)$, where $(t - t_0)$ is in days and t_0 is 0 hours UT on 1 January 1965.

terminator The boundary between the illuminated and unilluminated parts of the surface of a planet or moon.

tholin A high molecular weight mixture of organic compounds created in the laboratory in a hydrogen-rich atmosphere. This dark red substance is believed by some astronomers to be formed by photochemical reactions in the atmospheres of Jupiter and Titan.

Trojan asteroids Two families of asteroids which share the orbit of Jupiter, clustered around points 60° either side of the planet.

troposphere The lower layer of a planetary atmosphere where convection occurs.

UT Universal Time – the time system used for recording astronomical observations, approximately equivalent to the local mean time at 0° longitude (the Greenwich Meridian).

VLA Very Large Array – a radio telescope made up of 27 individual dishes located in New Mexico.

Voyager The name of two almost identical planetary probes launched to explore the outer solar system in 1977. Voyagers 1 and 2 both visited the jovian and saturnian systems. Voyager 2 went on to explore Uranus and Neptune, their moons and rings.

WFPC2 Wide Field Planetary Camera 2 – one of the imaging instruments on the Hubble Space Telescope, installed during the 1994 servicing mission.

World Wide Web A graphics-based system through which information can be made publicly available and accessible to users of the Internet electronic communications network.

WWW World Wide Web

The contributors

MICHAEL F. A'HEARN is a professor of astronomy at the University of Maryland. His principal research activities have been observational studies of comets at all wavelengths, from the ultraviolet to the radio, and he was one of the discoverers of gaseous jets in Comet Halley. He is also the manager of the Small Bodies Node of NASA's Planetary Data System.

CLARK R. CHAPMAN is a research scientist at the Planetary Science Institute, a division of San Juan Capistrano Research Institute, in Tucson, Arizona. He is a member of the Imaging Team of the *Galileo* Mission and of the Imaging/Spectroscopy Team of the Near Earth Asteroid Rendezvous Mission. An expert on asteroids and planetary cratering, Chapman was co-author (with D. Morrison) of the book *Cosmic Catastrophes* (Plenum, 1989). He was an avid observer of the planet Jupiter in the 1950s and 1960s, and wrote his Master's thesis on the circulation of Jupiter's clouds.

DAVID JEWITT is professor of astronomy at the University of Hawaii. He specializes in the study of comets and is currently investigating the properties of the trans-Neptunian objects using telescopes on Mauna Kea.

BRIAN G. MARSDEN is associate director for Planetary Sciences at the Harvard–Smithsonian Center for Astrophysics and director of the International Astronomical Union's Central Bureau for Astronomical Telegrams and Minor Planet Center. An authority on the orbits of comets and asteroids, he suggested within a week of its discovery that SL9 was orbiting Jupiter and concluded less than two months after its discovery, that the comet would collide with Jupiter on its next pass.

H. JAY MELOSH is at the Lunar and Planetary Laboratory of the University of Arizona, Tucson. His principal research interests are impact cratering, planetary tectonics, and the physics of earthquakes and landslides. His recent research includes studies of the giant impact origin of the Moon, the K/T impact that extinguished the dinosaurs, the ejection of rocks from their parent bodies and the breakup and collision with Jupiter of comet SL9

JACQUELINE MITTON is based in Cambridge, UK, where she went to study for her astronomy PhD. Since 1987, she has been a writer, editor and broadcaster specializing in astronomy for general audiences. She was appointed the first Public Relations Officer of the Royal Astronomal Society in 1989.

JOHN H. ROGERS is director of the Jupiter Section of the British Astronomical Association (BAA) and has written most of the BAA's Jupiter reports since 1972. He has a PhD in molecular biology but has always continued astronomical work as an amateur. He observes Jupiter with an historic 30-cm refractor at the University of Cambridge, England. By day, he is a researcher in molecular neurobiology and a lecturer in the Department of Physiology at the University of Cambridge.

CAROLYN S. SHOEMAKER began her work with the Palomar Asteroid and Comet Survey in 1982. Since that time she has discovered more than 800 asteroids, including 41 Earth approachers. She has discovered 32 comets, of which 15 are short-period and 17 are long-period comets. She is a research professor of astronomy at Northern Arizona University, a staff member at Lowell Observatory, and a volunteer with the US Geological Survey.

EUGENE (GENE) M. SHOEMAKER has recently retired, having been a geologist with the US Geological Survey. His long career in planetary science began with his studies of terrestrial impact craters and the geology of the Moon in the late 1950s. Most recently he has been the science team leader for the Clementine mission to the Moon. He was professor at the California Institute of Technology, 1969-1985, where he originated the Palomar Asteroid and Comet Survey. He is on the staff at Lowell Observatory and is a research professor of astronomy at Northern Arizona University. He has co-discovered 29 comets and many asteroids.

JOHN R. SPENCER has been fascinated by the Jupiter system since his teenage years. He is now a staff astronomer at Lowell Observatory in Flagstaff, Arizona, and studies the Galilean satellites of Jupiter, especially Io, using telescopes in Flagstaff and on Mauna Kea, Hawaii. He observed the comet crash from Cerro Tololo Observatory in Chile.

HAROLD (HAL) A. WEAVER is an astronomer at the Space Telescope Science Institute. His principal interest is cometary science, which he has been pursuing since 1979. In 1985–86 he made infrared observations of Halley's Comet from the Kuiper Airborne Observatory, which resulted in the first unambiguous detection of water in comets. For this work, he was awarded the NASA Medal for Exceptional Scientific Achievement in 1988. Lately his research has centered on HST observations of comets and he was Principal Investigator on the HST program to study comet P/Shoemaker–Levy 9.

KEVIN ZAHNLE works at NASA's Ames Research Center. Mostly he studies processes that were important in shaping the surface environment of planets and satellites when the solar system was young. He describes himself as an exobiologist. Earth is his favorite planet.

Index

Page numbers in *italics* signify a reference in a figure caption. Page numbers in **bold** signify a reference to an illustration.